ELECTRICAL, CONTROL ENGINEERING AND COMPUTER SCIENCE

PROCEEDINGS OF THE 2015 INTERNATIONAL CONFERENCE ON ELECTRICAL, CONTROL ENGINEERING AND COMPUTER SCIENCE (ECECS 2015), HONG KONG, 30–31 MAY 2015

Electrical, Control Engineering and Computer Science

Editor

Jian Liu
*School of Electrical and Information Engineering,
Wuhan Institute of Technology, Wuhan, China*

CRC Press
Taylor & Francis Group
Boca Raton   London   New York   Leiden

CRC Press is an imprint of the
Taylor & Francis Group, an **informa** business

A BALKEMA BOOK

CRC Press/Balkema is an imprint of the Taylor & Francis Group, an informa business

© 2016 Taylor & Francis Group, London, UK

Typeset by V Publishing Solutions Pvt Ltd., Chennai, India

Published by: CRC Press/Balkema
      P.O. Box 11320, 2301 EH Leiden, The Netherlands
      e-mail: Pub.NL@taylorandfrancis.com
      www.crcpress.com – www.taylorandfrancis.com

ISBN: 978-1-138-02937-8 (Hbk)
ISBN: 978-1-315-63882-9 (eBook PDF)

Table of contents

Communication and computer networks

Preface

The 2015 International Conference on Electrical, Control Engineering and Computer Science (ECECS2015) was successfully held in Hong Kong, May 30–31, 2015. The ECECS2015 was organized by the American Society of Science and Engineering (ASEE). The ASEE is a non-profit society for engineers and scientists, which was founded originally in 2009 and has been undergoing rapid expansion in the recent years. The ECECS2015 is co-sponsored by Chongqing University of Science and Technology, The Hong Kong Polytechnic University, MVGR College of Engineering, Babol University of Technology, Helwan University, Xi'an Jiaotong-Liverpool University, and North China Electrical Power University. The ECECS conference serves as an excellent platform for the engineering and science community to meet with each other and to exchange theories, ideas, techniques and experiences related to all aspects of electrical engineering, control engineering and computer science.

This book contains 39 revised and extended research articles, written by prominent researchers participating in the conference. Topics covered include electrical engineering, control engineering, communication and computer networks, and computer science. All accepted papers went through strict peer-reviewing by 2–4 expert referees and the overall acceptance rate was 38.9%. The papers have been selected for this book because of quality and the relevance to the conference. The organizing committee of ECECS2015 would like to express our sincere appreciations to all authors for their contributions to this book. We would like to extend our thanks to all the referees for their constructive comments on all papers; especially, we would like to thank to organizing committee for their hard working.

Prof. Jian Liu
General Chair of ECECS2015
Wuhan Institute of Technology

Electrical and control engineering

Electrical, Control Engineering and Computer Science – Liu (Ed.)
© 2016 Taylor & Francis Group, London, ISBN 978-1-138-02937-8

Structural design and motion simulation of a kind of AMT Clutch Actuator

Yanfang Zhang
Chongqing College of Electronic Engineering, China

ABSTRACT: Taking the AMT Clutch Actuator of an experimental platform of a kind of full hybrid electric vehicle as an example, we design the structure of the Clutch Actuator, select the type of motor for the Clutch Actuator, design and verify the joint strength of the actuator, and carry out 3D modeling and motion simulation of the Clutch Actuator in this paper.

Keywords: Clutch Actuator; structural design; verification; motion simulation

1 INTRODUCTION

As an important component of the mechanical transmission system of the vehicle, the clutch has functions of transmitting torques, absorbing shocks, resisting torsions, disengaging and jointing, etc. [1–2] The requirements for the design of the clutch are as follows: the clutch realizes the smooth jointing between the engine and the transmission making the vehicle start smoothly; breaks off the connection between the engine and the transmission system quickly reducing the impacts between the gears of the transmission and making the shifting easy; protects the transmission system from impact and destruction when suffering heavy dynamic loads during working. [3] In this paper, we design an AMT Clutch Actuator of which the driving mode is electronically controlled and electrically powered.

2 STRUCTURAL DESIGN OF THE CLUTCH ACTUATOR

The working process of the Clutch Actuator is: firstly, the ECU receives the shifting command according to the driver's intention and the real situation of the vehicle. Then the ECU controls the motor, and the motor drives the linear module, and the layer board on the linear module moves, driving the push-pull rod on it, which controls the piston of the main hydraulic cylinder which connects with a liquid storage cylinder to move towards the left or the right to adjust the pressure in the main hydraulic cylinder. The push of the piston in the main hydraulic cylinder makes the pressure in the oil passage on the right side of the main cylinder change, which produces a push-pull force to the clutch, and makes it meet the requirements for control. The structure of the actuator is shown in Figure 1.

3 THE DESIGN AND VERIFICATION OF THE CLUTCH CONTROL ACTUATOR

According to the experimental data of the experimental bench model, the complete disengaging time of the clutch should be less than 0.4 s, and the maximum pressure of the front-end hydraulic cylinder of the clutch is 0.8 MPa, so the design

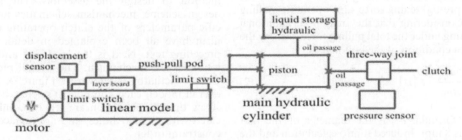

Figure 1. Samplified model for engine mounting.

of the actuator should meet the requirements of the maximum disengaging speed and the strength under the corresponding disengaging pressure. [4–5]

3.1 Selection of the motor type of the Clutch Actuator

In order to control the jointing speed and displacement of the clutch more accurately, we select the stepping motor. Subdivision control is adopted, so the developed control accuracy of the stroke displacement of the clutch can achieve 0.025 mm. According to the known conditions, the maximum pressure of the front-end hydraulic cylinder of the clutch is 0.8 MPa, and the diameter of the piston in the hydraulic cylinder of the actuator is 16 mm, so the following can be obtained based on this.

$$F_{max} = P_{max} \times A = 160N \qquad (1)$$

F_{max} refers to the maximum push-pull force of the push-pull rod of the actuator; A refers to the sectional area of the hydraulic cylinder.

The stroke of the clutch is $L = 30$ mm, the complete disengaging time of the clutch = 0.4 s. So the linear speed of the push-pull rod can be obtained. $v = L/t = 0.075$ m.

According to the calculations of the above formulas, the parameters of the driving motor of the Clutch actuator can be calculated. the power of the driving motor of the clutch: $P = F_{max} \times V$, which is calculated to be 12 W.

According to the calculated motor power, we select the Y200 L 1–2 stepping motor with a power of 30 W and a revolving speed of 2950 rpm. The front end of the push-pull rod is driven by a screw. Use the M10 × 0.5 metric fine screw thread for screw drive.

3.2 Design and verification of the actuator's connection strength

Use two M8 bolts initially for the threaded connection between the hydraulic cylinder and the guide plate. Bolts are being pulled. The bolts are under pre-tightening force and operating pulling force. Considering that they may need additional tightening under the total pulling force, thereby the strength condition of the bolt here is:

$$\sigma_e = \frac{5.2F_{max}}{\pi d_1^2} \leq [\sigma] \qquad (2)$$

In the formula, d_1 is the diameter of the bolt which is 8 mm. Induced it into calculation and the

conclusion is $\sigma_e = 4.14 \text{ MPa} < [\sigma] = 36 \text{ MPa}$, so the connection is safe.

Select M8 bolt initially for the threaded connection between the short axis and inhaul cable, and shear the bolt. The shear-resisting strength condition of the bolt is:

$$\tau = \frac{4F_{max}}{\pi d_0^2 m} \leq [\tau] \qquad (3)$$

In the formula, m is the bolt shank's shearing area, and $m = 2$; d_0 is the diameter of the bolt shank's shearing area, and $d_0 = 8$ mm. Induced it into calculation, and the conclusion is $\tau = 1.59 \text{ Mpa} \leq [\tau] = 45 \text{ Mpa}$, so the shear strength is safe. The extruding strength condition for the surface of the contact area between bolt shank and the wall of hole is:

$$\sigma_p = \frac{F_{max}}{d_0 h} \leq [\sigma_p] \qquad (4)$$

In the formula, h is the minimum height of the extruding area between bolt shank and the wall of hole, and $h = 6$ mm. Induce it into calculation, and the conclusion is $\sigma_p = 3.33 \text{ MPa} \leq [\sigma_p] = 90 \text{ MPa}$, so the extruding strength is safe.

The connection between the short axis and the layer board select two M6 bolts initially for connecting, and shear the bolts. The shear strength condition is as formula (3). In the formula, m = 1. Induce it into calculation, and the conclusion is $\tau = 5.66 \text{ MPa} \leq [\tau] = 45 \text{ MPa}$, Extruding strength condition is as formula (4). In the formula, h = 13.5 mm. Induce it into calculation, and the conclusion is $\sigma_p = 0.99 \text{ MPa} \leq [\sigma_p] = 90 \text{ Mpa}$, so the bolt connection here is safe.

4 3D MODELING AND SIMULATION OF THE CLUTCH ACTUATOR'S

4.1 Overall model of the Clutch Actuator

This article uses UG N6.0 for Clutch Actuator's 3D solid modeling, and adopts the top-down method to design the assembly. The specific design scheme, mechanism schematics and specific parameters of the clutch operating mechanism have all been explained in detail in the preceding text. Next is the modeling and simulation of the mechanism. The overall model of Clutch Actuator is shown in Figure 2. In this overall model, each component's screw bolt and screw thread connection are left out, only indicated with circular holes, and are replaced with constraint order.

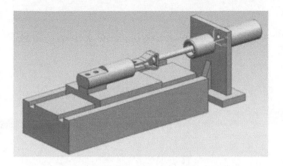

Figure 2. Overall model of Clutch Actuator.

4.2 Simulation analysis of the Clutch Actuator

Mechanism motion analysis module [motion], which is UG NX6.0 self-carried, provides mechanism simulation analysis and text production function. With a three-dimensional model of Clutch Actuator having been set up in the preceding text, the simulation analysis now follows.

Operate UG NX, open the set-up actuator mechanism model, and select [start]/[motion simulation] in the menu. Then single-click the right key in the assembly model of motion navigation, select [create a simulation] to create a simulation, and with a [motion-1] acquired, it enters the motion simulation mode.

First of all, assign link rods. Single-click the right key on [motion-1] to create link rods. With the link rods created, each independent space link rod has six degrees of freedom, and we need connect the link rods with kinematic pairs, which will form a certain constraint among the link rods, making the motion chain that is made up of the link rods have a definite motion, so as to create a mechanism.

Next, set up the sliding pair. Right-click on [motion-1] and then choose the sliding pair in the drop-down box of [create a new kinematic pair]. Later choose the connecting rod that needs to move. Using the line or the plane in its direction of motion as the chosen object can directly generate the kinematic pair, after which you can choose the direction of motion. Later, choose the truck mode

to be function in the driver dialog. Then choose the function manager to create a new function. Choose the motion function in the box after inserting. Create a STEP function and the body of this STEP function is STEP (time, 0, 0, 0.4, 60) + STEP (time, 0.4, 0, 1.4, 0) + STEP (time, 1.4, 0, 1.8, −60).

When the connecting rod and the kinematic pair are available, create a new budget scheme. What needs to be noticed is that the number of steps should be set to 1000 steps, which is beneficial to the observation of the simulation result. When the system settlement is finished, the simulation animation is available, verifying the rationality of the clutch control actuator.

5 CONCLUSION

This paper introduces a AMT Clutch Actuator applied to full hybrid electric vehicle experimental platforms. It carries out design calculation and strength check and gets the basic structure and dimensions. It uses UG NX6.0 to design the structure and generates the three-dimensional model of AMT Clutch Actuator and conducts motion simulation in the end, verifying the rationality of this clutch control actuator.

REFERENCES

[1] Wang Xiaochuang, Tang Guangdi et al. Research on A Kind of Hybrid Power Motor Automatic Clutch Actuator [J]. Mechanical Drive, 2013.
[2] Lun, Cheng Xiusheng, Ge Anlin, et al. Clutch Control in the Process of AMT Shifting [J]. Automobile Technology, 2006.
[3] Tian Ye. AMT Automatic Shifting Speed Control System Design [J]. Shanxi Electronic Technology, 2009.
[4] Zhang Hui, Liu Zhenjun, Qin Datong. Heavy Duty Motor AMT Hydraulic Shifting Executing Agency Analysis and Design [J]. Hydraulic and Pneumatic, 2007.
[5] Jin Long. AMT Electric Clutch Actuator Dynamic Property Simulation Research [D]. Jilin University, 2007.

Electrical, Control Engineering and Computer Science – Liu (Ed.)
© 2016 Taylor & Francis Group, London, ISBN 978-1-138-02937-8

Comparative study of electrical characteristics in southern, central, and northern areas of Jiangsu province

Cheng-chao Jiang, Pei-hong Wang & Yong-sheng Hao
Key Laboratory of Energy Thermal Conversion and Control of Ministry of Education, Southeast University, Nanjing, Jiangsu Province, China

ABSTRACT: This paper selects Nanjing, Changzhou, and Xuzhou as the representatives of cities in central, southern, and northern areas of Jiangsu province, respectively. The purpose of research is to discover electrical characteristics in different areas in Jiangsu province by comparing marketing data of Nanjing, Changzhou, and Xuzhou. Matlab is adapted to process large amounts of marketing data. The result shows that the average capacity in Nanjing and Changzhou is larger than that in Xuzhou. Similarly, the line loss rate in Nanjing and Changzhou is lower than that in Xuzhou.

Keywords: comparison; marketing data; line loss rate; average capacity

1   INTRODUCTION

Due to different economic development levels and industrial structures among regions, the electrical characteristics of each region are different. Taken the Jiangsu province as an example, area in southern Jiangsu is much more economically developed while the industrial base of northern Jiangsu area is relatively weak and the first industry still occupies a considerable proportion in industrial structure. Area in central Jiangsu is in between. In this paper, Nanjing, Changzhou, and Xuzhou are selected as study samples. Nanjing is the capital city of Jiangsu province and also the typical city in central Jiangsu area; Changzhou is the typical city in southern Jiangsu area; Xuzhou is the typical city in northern Jiangsu area. The differences of electrical characteristics among these three cities reflect the differences of that in different areas of Jiangsu province, even China. Therefore, a comparative study of electrical characteristics among Nanjing, Changzhou, and Xuzhou is significant in guiding the management of electricity in different cities. When it comes to the electrical characteristics, Wang explored the electrical characteristics between the different region markets in America by multifractal approaches (Wang 2013). However, there are more electrical characteristics, such as line loss rate and average capacity and so on, and differences of that among typical cities in China still worth studying.

This paper analyzes marketing data of December 2015 from 94198 low-voltage districts in Nanjing, Changzhou, and Xuzhou. Marketing data of power company includes parameters of every district like total number of electricity users, total capacity, number of residential electricity users, number of non-residential electricity users, total capacity of residential users, total capacity of non-residential users, and line loss rate. These parameters are first parameters, which are acquired directly by data system of power company. There are second parameters derived from first parameters, including residential capacity ratio and average capacity. The second parameters are based on the following definitions:

$$Rcr = \frac{Rc}{(Rc + Nc)} \qquad (1)$$

In Eq. 1, Rcr stands for the residential capacity ratio, Rc for the total capacity of residential users, and Nc for the total capacity of non-residential users.

$$Dc = \frac{Rc}{Nr} \qquad (2)$$

In Eq. 2, Dc stands for the average capacity, Rc for the total capacity of residential users, and Nr for the number of residential electricity users.

Since residential capacity ratio, average capacity, and line loss rate can reflect, respectively, composition of electricity users, level of electricity consumption, and comprehensive management level in electricity, these three parameters are employed to analyze electrical characteristics of three cities.

2 COMPARATIVE ANALYSIS OF ELECTRICAL CHARACTERISTICS OF THREE CITIES BASED ON AVERAGE CAPACITY

According to the marketing data from urban and rural power grid in three cities, scatter plots of average capacity-residential capacity ratio are plotted by use of matlab, as shown in Figure 1.

Figure 1 depicts the distribution of average capacity with residential capacity ratio. Each spot denotes a district. It is evident from Figure 1 that points congregate in some certain average capacities and the distribution of points differs among cities. The average capacity of Nanjing urban districts mainly congregates in 4 kVA, 8 kVA, and 12 kVA while that of rural districts mainly congregates in 4 kVA, 8 kVA; average capacity of

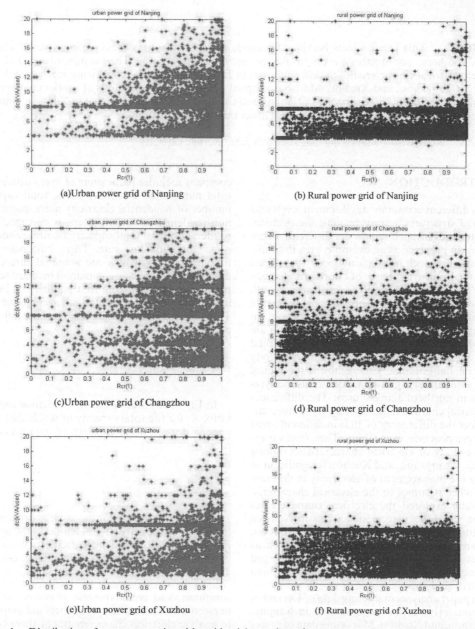

(a)Urban power grid of Nanjing

(b) Rural power grid of Nanjing

(c)Urban power grid of Changzhou

(d) Rural power grid of Changzhou

(e)Urban power grid of Xuzhou

(f) Rural power grid of Xuzhou

Figure 1.   Distribution of average capacity with residential capacity ratio.

Changzhou urban districts mainly congregates in 2 kVA, 4 kVA, 8 kVA, 12 kVA while that of rural districts mainly congregates in 4 kVA, 8 kVA, 12 kVA; average capacity of Xuzhou urban districts mainly congregates in 2 kVA, 8 kVA while that of rural districts mainly congregates in the section of 1~8 kVA. Urban districts of Nanjing and Changzhou have a relatively higher level of average capacity with lots of districts above 8 kVA, the whole, compared with Xuzhou. This indicates that Nanjing and Changzhou have much higher level of electricity consumption and standard of living than Changzhou. In addition, comparing the points of Nanjing and Changzhou, a significant difference between two cities is that Nanjing urban districts have a far greater proportion of pure-residential

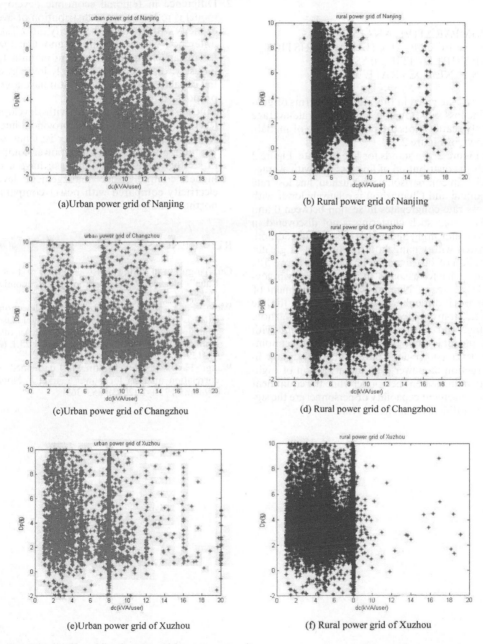

(a)Urban power grid of Nanjing

(b) Rural power grid of Nanjing

(c)Urban power grid of Changzhou

(d) Rural power grid of Changzhou

(e)Urban power grid of Xuzhou

(f) Rural power grid of Xuzhou

Figure 2.   Distribution of line loss rate with average capacity.

9

districts whose residential capacity ratio are above 0.85 than Changzhou. This indicates that industries are widespread in Changzhou. This difference between Nanjing and Changzhou also mirrors the difference in economic development model, which industry in Changzhou makes more contribution to local economy. The difference among rural power grids is basically in line with such among urban power grids.

3 COMPARATIVE ANALYSIS OF ELECTRICAL CHARACTERISTICS OF THREE CITIES BASING ON LINE LOSS RATE

Line loss rate given in this paper is in terms of general line loss rate (Gu 2008). Line loss rate-average capacity scatter plots are plotted by use of matlab, as shown in Figure 2.

In Figure 2, Dp stands for line loss rate. Figure 2 depicts the distribution of line loss rate with average capacity. In comparison with Xuzhou, line loss rate of Nanjing and Changzhou is relatively lower with line loss rate congregates in section between 0 and 4%. However, such clustering is not discovered in Xuzhou, indicating higher line loss rate. In addition, compared with Changzhou, Nanjing has a greater proportion of districts with line loss rate above 5%. The most likely reason is that old downtown covers a large area of Nanjing where refurbishment of power grid is difficult to implement. Overall, the line loss rate from low to high in order is: Changzhou, Nanjing, and Xuzhou, which is consistent with development level of cities. This is because economically developed regions have a huge advantage in construction of power grid and attraction of high-quality management personnel while the equipment and management capability of personnel are the significant influences on line loss rate.

4 CONCLUSIONS

1. The level of average capacity reflects regional living standard and level of electricity consumption. For Jiangsu province, areas in southern and central Jiangsu have a much higher level of average capacity with lots of urban districts above 8 kVA while most districts in Xuzhou are below 8 kVA.
2. Difference in regional economic development model is reflected in the distribution of average capacity with residential capacity ratio. Take the difference between Nanjing and Changzhou, for instance, due to the strong manufacturing industry; districts with relatively lower residential capacity ratio in Changzhou have a greater proportion than Nanjing.
3. Line loss rate is consistent with development level of cities. For Jiangsu province, line loss rate from low to high in order is: Changzhou, Nanjing, and Xuzhou. Thus, power companies in southern and central Jiangsu have a much higher comprehensive management level in electricity compared with power companies in northern Jiangsu.

REFERENCES

Gu, Yu-gui, et al. "Study on Calculation of Line Loss Rate." Electric Power Technologic Economics, 20 (2008): 34–37.

Wang, Fang, et al. "Cross-correlation detection and analysis for California's electricity market based on analogous multifractal analysis." Chaos: An Interdisciplinary Journal of Nonlinear Science 23.1 (2013): 013129.

Wang, Fang, et al. "Multifractal detrended cross-correlation analysis for power markets." Nonlinear Dynamics 72.1–2 (2013): 353–363.

Electrical, Control Engineering and Computer Science – Liu (Ed.)
© 2016 Taylor & Francis Group, London, ISBN 978-1-138-02937-8

The change rule of line loss rate based on the marketing data of the stable districts in Jiangsu province

Jin-ling Xu, Yun-feng Zou & Yue-ping Kong
Jiangsu Electric Power Research Institute, Nanjing, Jiangsu Province, China

ABSTRACT: This paper studies the major influence factors as well as the response rule of the low-voltage district line loss based on the marketing data, providing methods support for line loss forecast and abnormal lines loss assessment. This paper uses the marketing data sample of the stable low-voltage district in the Jiangsu province, by abnormal data filter, capacity classification in per household and regression modeling of line loss rate, to acquire the response rule of the line loss of various typical samples in stable district to the user number and load rate, which could be a reference for line loss forecast and abnormal assessment.

Keywords: line loss of low-voltage district; classification of district; load rate; total user number

1 INTRODUCTION

Line loss is a very important technical-economic indicator for power grid enterprise, which reflects the planning and design of power grid, production technology, and operation management level[1]. Line loss management in low-voltage district is the key to line loss management for power grid enterprise, which involves planning, operation, maintenance, measure, electric charge, electricity inspection, and many other professions, covering more than 95% of the electricity users.

For a long time, although all the power supply enterprises attach great importance to the line loss management in low-voltage district, there still exist many problems in power marketing management and equipment management, such as inaccurate household relationship, measuring device faults, low meter reading quality, electric-theft, long power supply line, three-phase unbalance, leakage of electricity, poor connector contact, and so on. These problems are mixed with each other, which makes it difficult to manage and lower the line loss in low-voltage district.

With the construction and application of smart meters and electricity information collection system, remote on time automatic meter reading of low voltage area metering points and user metering points can be realized, which greatly improves the real time and accuracy of line loss management in low-voltage district and effectively promotes the rise of level power grid operation and management. State Grid Jiangsu Electric Power Company started to manage the line loss in low-voltage district based

on data gathered from power customer's information, and the number of districts of which line loss rate is from −1% to 5% and increase to 89.1% after nearly two years of efforts. The economic benefit is remarkable.

In order to further improve the level of line loss management, a reasonable line loss rate reference for each district and online monitoring is needed so that abnormal area can be discovered in time. Then reasons can be analyzed and management problems can be solved without delay. But theoretical line loss calculation is very difficult because complex low-voltage district branch circuit, diverse components, incomplete equipment data and real time need cannot be met. Therefore, how to quickly calculate the reasonable district line loss has become a hot area of current research on low-voltage district line loss.

In this essay, the response rules between district average load rate, total number of user, and district line loss rate will be studied by adopting the method of mathematical modeling and statistical theory based on the causes of line loss to provide a new solution to reasonable line loss calculation and abnormal evaluation research.

2 ANALYSIS OF DISTRICT LINE LOSS CAUSES

In [2], it is pointed out that low-voltage district line loss is usually divided into two types, the first type is fixed loss $\Delta p1$, which is produced once the electrical device, is electrified; in general, it does not vary with the change in load. The second type is the change loss $\Delta p2$, which is in direct proportion to the square of the current. The loss of the low-voltage in district is bigger if the current is larger; it varies with the change of load. It is mainly produced by the resistance of the circuit itself.

$$\Delta p2 = \int_0^t I^2(t)Rdt \tag{1}$$

$$P = \int_0^t I(t)Udt \tag{2}$$

P is power supply volume.

$$\rho = \frac{\Delta p1 + \Delta p2}{P} = \frac{\Delta p1 + \int_0^t I^2(t)Rdt}{\int_0^t I(t)Udt} \tag{3}$$

ρ is line loss rate. Supposing that the current $I(t)$ is the constant I which represents the average current. (3) is simplified as follows:

$$\rho = \left(c_0 I + \frac{\Delta p1}{P}\right) \times 100\% \tag{4}$$

where c_0 is a constant. Fixed line loss $\Delta p1$ can be achieved through the electric power system information collection system. This paper studies line loss rate which gets rid of the fixed line loss, defined as Dp.

$$Dp = c_0 I \times 100\% \tag{5}$$

It can be seen from (5) that line loss rate is primarily related to current, and the current is closely associated with load. The amount of total user number can reflect the power supply radius to a certain degree, and the power supply radius determines the resistor R. This paper mainly studies the influence of load rate and the total user number on the line loss rate. The formula of load rate is:

$$Rl = P/(Ct \times 31 \times 24) \tag{6}$$

where 31 means the days of a month and 24 means hours of a day. Ct is the transformer capacitor.

3 DATA FILTER AND CLASSIFICATION

This thesis uses the marketing data of the stable district of which daily line loss rate does not fluctuate a lot in December in 2014 in Jiangsu as data analysis. The data include 160932 districts, eight variables that are total user number N, residential number Nr, residential capacity Cr, non-residential capacity Cc, transformer capacitor Ct, power supply volume P, power sale volume Ps, and fixed loss $\Delta p1$.

1. Data filter is based on daily power consumption. First, daily power consumption per household is defined by the following formula:

$$de = P/(N \times 310) \quad unit: \text{kW/h} \tag{7}$$

The daily power consumption threshold can be determined by statistical distribution further then improper data should be deleted based on Figure 1. According to the sample data of the 160932 stable districts, the lower limit of the daily power consumption threshold is 1.2 kW.h. the upper limit of the threshold is 20 kW.h. 131893 districts are qualified.

2. Data classification is based on line loss rate. According to Figure 2 and Enterprise Management standard, the lower limit and upper limit of the line loss rate threshold are −1% and 5%, respectively. Among the 160932 districts samples in Jiangsu, 147238 districts are qualified.

3. Users in district are divided into residential users and non-residential users, the characteristic of power utilization differs a lot. The proportion

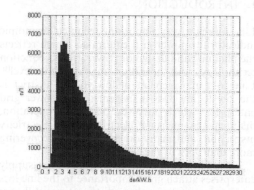

Figure 1. Distribution rule of power consumption per household—number of districts.

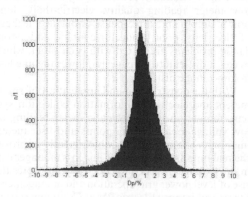

Figure 2. Distribution rule of line loss rate—number of districts.

of residential capacitor is defined by the following formula:

$$Rcr = Cc/(Cc + Cr) \qquad (8)$$

In this paper, the district with the feature of $Rcr > 0.6$ is defined as the resident district. In view of the complexity and variety of the mixed and non-resident district, this paper focuses on the resident district.

4. Many factors influence the line loss. For example, the existence of dual structure of economic development of urban and rural as well as the difference of electrical usage between urban and rural residents can result in the difference between amounts of line loss. Meanwhile, differences of power supply radius, distribution technology, and route planning can influence the line loss to a large degree. Therefore, the reasonable data can be classified as urban and rural by difference of regions[3].

5. Capacity of per household is defined by the following formula:

$$dc = Cr/Nr \; unit: \; kV.A \qquad (9)$$

It can be seen from Figure 3 that capacitor per household mainly falls around 4 kV.A, 8 kV.A, and 12 kV.A. Therefore, this paper assumes that $2 < = dc < 6$kV.A is low-type district, $6 < = dc < 10$ kV.A is middle-type district, and $10 < = dc < = 14$ kV.A is high-type district.

4 RESULTS AND ANALYSIS OF CLASSIFICATION[4]

By the elimination standard and classification conditions, the results are divided into 6 classes: they are 3024 low-type districts in urban areas, 16417 middle-type districts in urban areas, 2879 high-type districts in urban areas, 31169 low-type districts in rural areas, 47248 middle-type districts in rural areas, and 1846 high-type districts in rural areas, and the total amount is 102768.

The load rate variable is taken as an example to reflect the relationship between line loss rate and variable more effectively. First, sample space is sampled when the load rate is 0 and the step size is 0.005, if the districts amount is less than 100 in the sample space, the step size is changed to 0.01,

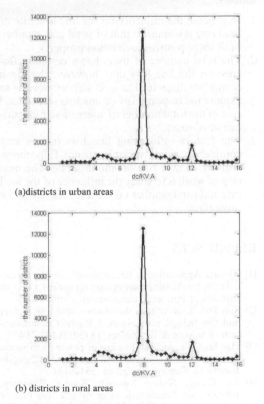

(a)districts in urban areas

(b) districts in rural areas

Figure 3. Distribution of capacitor per household—amount of transformer district.

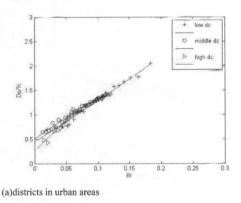

(a)districts in urban areas

(b) districts in rural areas

Figure 4. Fitting relationship figure of line loss rate—load rate distribution.

and so on. Dividing the classified data into sample spaces of C1, C2,...Cn, calculating the average load rate and line loss rate of the n sample spaces to replace the load rate and line loss rate of the sample space, and then, fitting analysis of polynomial can be conducted.

4.1 Relationship between load rate and line loss rate

It can be seen from Figure 4(a) that among the stable districts in Jiangsu, change interval of load rate in lower type residents is much larger than that of high-type residents, and change interval of load rate in middle-type residents is between the two above. Line loss rate is in direct proportion to load rate. The Figure 4(b) shows that trend of line loss rate of stable districts in rural areas is familiar to that of urban areas; the load rate of lower type resident district is apparently less than that of urban areas. The reason is that living standard in urban areas is better than rural areas.

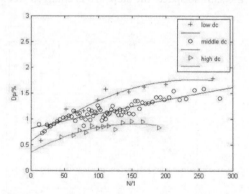

(a)districts in urban areas

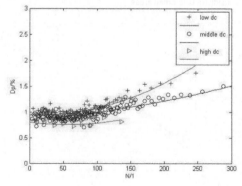

(b) districts in rural areas

Figure 5.   Fitting relationship figure of line loss rate—total user number.

4.2 Relationship between total user number and line loss rate

It can be seen from Figure 5(a), (b) that, on the whole, line loss rate increases with the rise of the total user number. In general, line loss rate is smaller if capacity per household is bigger when the total number of household is the same. When N < 100, in rural area, line loss rate does not fluctuate a lot, however, when N > 100, line loss rate grows with the rise of the total user number.

By comparing the influence on the line loss rate of load rate and total user number, it can be found that the influence of load rate on line loss rate is more apparent.

5   CONCLUSIONS

This paper analyzes the change rule of line loss rate in view of marketing data, the variables which are not directly related to the line loss rate; therefore, it is difficult to find the relationship between the sample variables and line loss rate. By exploratory research on the data sample of the stable district in Jiangsu, conclusions in this paper are as follows:

1. In general, the influence on the line loss rate of load rate is similar to that of total user number, both show positive correlation property.
2. The total number of users has a certain influence on the line loss rate; however, the mean points are discrete. The next step of work is to explore the response law of line loss rate to load rate in the total number of users of size classification of cases.
3. The factors influencing line loss rate is very complex, a single variable cannot fully represent the change rule of line loss rate. The next step of work is to study the influence of the load rate and total number on the line loss rate at the same time.

REFERENCES

[1] Zhu Jie, Application of data mining in line losses calculation for electric power marketing system, D. Xi'an Institute of Post and Telecommunications, 2008.
[2] Sun Hui, Low voltage distribution line loss factors and loss reduction measures, J. Public Communication of Science & Technology. 15 (2012) 103–104.
[3] Hao ling-Xia, Analysis line loss of power distribution network and line loss reduction measures, J. Chengshi Jianshe yu Shangye Wangdian. 35 (2012).
[4] Xu Qiang, Shang Xue-bin, Kuang Wei-hong, Research on Benchmark value of line loss rate of typical district, J. China High-Tech Enterprises. 316 (2015) 31–35.

Electrical, Control Engineering and Computer Science – Liu (Ed.)
© 2016 Taylor & Francis Group, London, ISBN 978-1-138-02937-8

A comparative study of mode-space transform and array interpolation for DOA estimation of coherent signals in fourth-order cumulant domain

Menghan Liu
School of Electronic Engineering, University of Electronic Science and Technology of China, Chengdu, Sichuan, China
Science and Technology on Electronic Information Control Laboratory, Chengdu, China

Shangwei Gao
Science and Technology on Electronic Information Control Laboratory, Chengdu, China

Jihao Yin
School of Electronic Engineering, University of Electronic Science and Technology of China, Chengdu, Sichuan, China

ABSTRACT: This paper focuses on the Direction of Arrival (DOA) estimation of coherent signals based on a Uniform Circular Array (UCA) in the fourth-order cumulant Domain. The mode-space transform and array interpolation transform, which both can transform the raw data received by a UCA into a virtual Uniform Linear Array (ULA) model, are used as a preprocessing technique in a comparative way. Using two virtual ULA models given by these two preprocessing methods, we construct a fourth-order cumulant matrix, together with the well-known spatial smoothing, to obtain two DOA estimators for coherent signals. The performance comparison of the two DOA estimators are done by some theoretical analysis and simulations.

Keywords: array preprocessing; coherent signals; DOA estimation; fourth-order cumulant

1 INTRODUCTION

The Direction of Arrival (DOA) estimation of coherent signals is an important problem in communication and radar, since multi-path propagation or spot jamming often occurs in practice. The well-known spatial-smoothing technique has proved to be effective for the DOA estimation of coherent signals, but it usually requires a Uniform Linear (ULA) array with the Vandermonde structure. On the other hand, Uniform Circular Arrays (UCAs) have been widely used in practice due to their isotropic DOA estimation performance in the sense of Cramer-Rao bound. The spatial-smoothing cannot be directly applied to the array data model based on a UCA, and some preprocessing technique is needed.

In this paper, we focus on two preprocessing techniques: the mode-space transform and array interpolation transform, which transform the raw data received by a UCA into two virtual uniform linear array models, respectively. Based on these two virtual ULA models, we construct

a fourth-order cumulant matrix, together with the well-known spatial smoothing, to obtain two DOA estimators for coherent signals. Fourth-order cumulants usually have a few advantages over second-order statistics in array signal processing, such as the (asymptotical) insensitiveness to Gaussian noise that can be used to suppress Gaussian noise without knowing or estimating the noise covariance and the ability of array aperture extension that can improve the DOA estimation resolution. It should be pointed out that signals are required to have non-zero fourth-order cumulants when fourth-order cumulants are used in the DOA estimation. The performance comparisons of the two DOA estimators are done by some theoretical analysis and simulations.

2 TWO VIRTUAL DATA MODEL GIVEN BY TWO PREPROCESSING TECHNIQUES

Consider L narrow coherent signals, corrupted by Gaussian noise, impinge from $\theta = [\theta_1, \theta_2, \ldots, \theta_L]$ on a UCA with M sensors and a radius r. The array data $x(t)$ is

$$x(t) = A(\theta)s(t) + e(t), \quad t = 1, 2, ..., N, \qquad (1)$$

where $A(\theta) = [a(\theta_1), a(\theta_2), ..., a(\theta_l), ..., a(\theta_L)]$, and $a(\theta_l)$ is the lth signal's array steering vector with the mth element $a_m(\theta_l) = \exp[-j(2\pi r/\lambda)\cos(\theta_l - 2\pi(m-1)/M)]$ where λ is the wave length of signals. $s(t) = [s_1(t), s_2(t), ..., s_L(t)]^T$ is the signal vector and $(\cdot)^T$ is the transpose operation. Without loss of generality, we choose $s_1(t)$ as the reference for the coherent signals so that $s_l(t) = \rho_l s_1(t)$ where ρ_l is the lth coherent coefficient and satisfies $|\rho_l| = 1$ ($l = 2, 3, ..., L$). $e(t)$ is the Gaussian noise vector. N is the number of snapshots.

Since $a(\theta_l)$ in (1) has no Vandermonde structure and the spatial smoothing cannot be directly applied, the data model (1) should first be transformed into a virtual ULA-based data model, by some preprocessing technique that usually is a linear transform. Denote this linear transform by T, which can provide us with a virtual array having the Vandermonde structure. Applying T to both sides of (1) yields

$$y(t) = Tx(t) = \tilde{A}(\theta)s(t) + Te(t), \quad t = 1, 2, ..., N, \quad (2)$$

where $\tilde{A}(\theta) = TA(\theta)$ is the virtual array manifold matrix whose lth column corresponds to the virtual array steering vector of the lth signal. In the literature, there are several methods for the design of T. Herein, we focus on two preprocessing techniques: the mode-space transform and array interpolation transform. For clarity, below we denote the mode-space transform and array interpolation transform as T_I and T_{II}, respectively.

2.1 Mode-space transform

According to the last one of the references, the well-known mode-space transform is derived mainly based on the periodicity of the electromagnetic field excited on a UCA and then through the Fourier transform analysis, which is given by

$$T_I = \frac{1}{M}J^{-1}F^H, \qquad (3)$$

where

$$J = \begin{bmatrix} j^{-k}J_h(-2\pi r/\lambda) & 0 & 0 \\ 0 & \ddots & 0 \\ 0 & 0 & j^k J_k(-2\pi r/\lambda) \end{bmatrix},$$

$$F = \begin{bmatrix} \exp(-j2\pi K0/M) & \exp(-j2\pi K/M) & \cdots & \exp(-j2\pi K(M-1)/M) \\ \exp(-j2\pi(K-1)0/M) & \exp(-j2\pi(K-1)/M) & \cdots & \exp(-j2\pi(K-1)(M-1)/M) \\ \vdots & \vdots & \cdots & \vdots \\ \exp(j2\pi K0/M) & \exp(j2\pi K/M) & \cdots & \exp(j2\pi K(M-1)/M) \end{bmatrix},$$

$J_k(\cdot)$ is the kth-order Bessel function, K is a positive integer satisfying $2K < M$, $(\cdot)^{-1}$ is the matrix inverse operation, $(\cdot)^H$ is the conjugate and transpose operation. When applying T_I to the data model (1), the virtual array manifold matrix is

$$\tilde{A}_I(\theta) = T_I A(\theta)$$
$$= [\tilde{a}_I(\theta_1), \tilde{a}_I(\theta_2), ..., \tilde{a}_I(\theta_l), ..., \tilde{a}_I(\theta_L)]$$
$$= \begin{bmatrix} \exp(-jK\theta_1) & \cdots & \exp(-jK\theta_L) \\ \vdots & \cdots & \vdots \\ \exp(jK\theta_1) & \cdots & \exp(jK\theta_L) \end{bmatrix}. \quad (4)$$

2.2 Array interpolation transform

The first step for designing an array interpolation transform is to determine the form of the virtual array that we assign to the given real array. As is done in the literature, we assume the form of the virtual array is a uniform linear array of M' sensor elements with the element interval of $\lambda/2$. The second step is to divide the field of view of the array into a few sectors and select out one or several sectors that cover the real DOAs θ. Then we determine a grid of angles within the selected sectors and figure out the real array manifold matrix and the virtual array manifold matrix on the grid. At last the array interpolation transform matrix is designed by the real and virtual array manifold matrices.

For example, we divide the whole field of view $(0°, -360°)$ into a few sectors (let us say each sector has an angle range of 50° or 60°) and select out the sector that covers the real DOAs θ and is denoted by Θ. Define a grid within Θ with the step $\Delta\theta$ (The step is usually set to 0.5° or 1°.): $\Theta = [\Theta_0, \Theta_0 + \Delta\theta, \Theta_0 + 2\Delta\theta, ..., \Theta_1]$, where Θ_0 and Θ_1 are two endpoints of Θ. Denote the real array manifold matrix and the virtual array manifold matrix on the grid as $A(\Theta) = [a(\Theta_0), a(\Theta_0 + \Delta\theta), ..., a(\Theta_1)]$ and $\tilde{A}(\Theta) = [\tilde{a}(\Theta_0), \tilde{a}(\Theta_0 + \Delta\theta), ..., \tilde{a}(\Theta_1)]$, respectively. Then the array interpolation transform matrix can be designed by

$$T_{II} = (B^H B)^{-1/2} B^H, \qquad (5)$$

where $B = (A(\Theta)A^H(\Theta))^{-1}A(\Theta)\tilde{A}^H(\Theta)$. When applying T_{II} to the data model (1), the real array manifold matrix $A(\theta)$ is in principle mapped into the column span of the virtual array manifold matrix:

$$\tilde{A}_{II}(\theta) = [\tilde{a}_{II}(\theta_1), \tilde{a}_{II}(\theta_2), ..., \tilde{a}_{II}(\theta_l), ..., \tilde{a}_{II}(\theta_L)]$$

$$= \begin{bmatrix} \exp(j(2\pi d/\lambda)K'\sin(\theta_1)) & \cdots \\ \exp(j(2\pi d/\lambda)(K'-1)\sin(\theta_1)) & \cdots \\ \vdots & \vdots \\ \exp(-j(2\pi d/\lambda)(K'-1)\sin(\theta_1)) & \cdots \\ \exp(-j(2\pi d/\lambda)K'\sin(\theta_1)) & \cdots \end{bmatrix}$$

$$\begin{matrix} \cdots & \exp(j(2\pi d/\lambda)K'\sin(\theta_L)) \\ \cdots & \exp(j(2\pi d/\lambda)(K'-1)\sin(\theta_L)) \\ \vdots & \vdots \\ \cdots & \exp(-j(2\pi d/\lambda)(K'-1)\sin(\theta_L)) \\ \cdots & \exp(-j(2\pi d/\lambda)K'\sin(\theta_L)) \end{matrix} \Bigg], \quad (6)$$

where $K' = (M'-1)/2$ or $K' = M'/2$, depending on M' is even or odd, and $d = \lambda/2$. Note that in view of the symmetric structure of a UCA, we specially endow the virtual ULA in the array interpolation transform with a center symmetry.

2.3 Comparison of two preprocessing techniques

It is clear that $\tilde{A}_I(\theta)$ given by the mode-space transform and $\tilde{A}_{II}(\theta)$ given by the array interpolation transform both have the Vandermonde structure and therefore satisfy the application condition of the spatial smoothing technique. However, it is also obvious that the two virtual arrays corresponding to $\tilde{A}_I(\theta)$ and $\tilde{A}_{II}(\theta)$, respectively, are quite different, since $\tilde{a}_I(\theta)$ is related to θ via the expression $jk\theta$ ($k = -K, ..., K$) while $\tilde{a}_{II}(\theta)$ is related to θ via the expression $j2\pi dk'\sin(\theta)/\lambda$ ($k' = -K', ..., K'$). In general, the array interpolation transform can be used for many array structures including UCAs and nonuniform linear arrays, while the mode-space transform can only be applied to UCAs as it is derived from the periodicity of the electromagnetic field excited on a UCA. On the other hand, the array interpolation transform requires a priori that the selected sector Θ should cover the real DOAs θ. This a priori knowledge can be provided by some beamforming technique in practice. In contrast, the mode-space transform needs no such a priori. In addition, since the mode-space transform and the array interpolation transform are both linear transforms, the noise term $Te(t)$ in (2) is still Gaussian and can be suppressed by fourth-order cumulants in principle. Therefore, the noise term in (2) is ignored below with no ambiguity.

3 THE CONSTRUCTION OF CUMULANT MATRIX FOR THE DOA ESTIMATION

Before constructing cumulant matrix, we summarize briefly the process of using fourth-order cumulants, along with spatial smoothing, to handle our DOA estimation problem based on the two virtual data models given by the two preprocessing techniques. We first rewrite (2) in another form to facilitate the analysis and derivations below. Since $s_l(t) = \rho_l s_1(t)$, $l = 2, 3, ..., L$, (2) can be reformulated as

$$y(t) = \tilde{A}(\theta)ds_1(t), \quad t = 1, 2, ..., N, \quad (7)$$

where $d = [1, \rho_2, \rho_3, ..., \rho_L]^T$. In order to apply spatial smoothing, we then divide the vector $y(t)$ into Q subvectors $y_q(t) = \tilde{A}_q(\theta)ds_1(t)$, $q = 1, 2, ..., Q$, where the length of $y_q(t)$ is P and $\tilde{A}_q(\theta)$ is the matrix formed by the first $(P+q-1)$ rows of $\tilde{A}(\theta)$. For $y_q(t)$, $q = 1, 2, ..., Q$, Q fourth-order cumulant matrices are constructed. At last, we average these Q cumulant matrices and use MUSIC DOA estimation method to obtain coherent signals' DOA estimates. Given below is the construction of a cumulant matrix.

The cumulant matrices, denoted by C_q for $y_q(t)$, are constructed as follows:

$$C_q = \begin{bmatrix} \tilde{c}_{y_q}(1,1) & \tilde{c}_{y_q}(1,2) & \cdots & \tilde{c}_{y_q}(1,P) \\ \tilde{c}_{y_q}(2,1) & \tilde{c}_{y_q}(2,2) & \cdots & \tilde{c}_{y_q}(2,P) \\ \vdots & \vdots & \vdots & \vdots \\ \tilde{c}_{y_q}(P,1) & \tilde{c}_{y_q}(P,1) & \cdots & \tilde{c}_{y_q}(P,P) \end{bmatrix}, \quad (8)$$

where the scalar $\tilde{c}_{y_q}(m,n)$ is

$$\tilde{c}_{y_q}(m,n)$$

$$= cum(y_{qm}, y^*_{qn}, y_{K+1}, y^*_{K+1})$$

$$= E\{y_{qm}y^*_{qn}y_{K+1}y^*_{K+1}\} - E\{y_{qm}y_{K+1}\}E\{y^*_{qn}y^*_{K+1}\}$$

$$\quad - E\{y_{qm}y^*_{K+1}\}E\{y^*_{qn}y_{K+1}\}$$

$$\quad - E\{y_{qm}y^*_{qn}\}E\{y_{K+1}y^*_{K+1}\}.$$

and $cum(y_{qm}, y^*_{qn}, y_{K+1}, y^*_{K+1})$ is the fourth-order cumulant of $(y_{qm}, y^*_{qn}, y_{K+1}, y^*_{K+1})$. y_{qm} is the mth element of $y_q(t)$ and y_{K+1} is the $(K+1)$th element of $y(t)$.

C_q in (8) satisfies

$$C_q = \tilde{A}_1(\theta)D_q\tilde{A}_1^H(\theta), \quad q = 1, 2, ..., Q, \quad (9)$$

where

$$D_q = C_s \left| 1 + \sum_{l=2}^{L} \rho_l \right|^2 (\boldsymbol{\Phi}^q d)(\boldsymbol{\Phi}^q d)^H,$$

$$
\begin{aligned}
C_s &= cum(s_1, s_1^*, s_1, s_1^*) \\
&= E\{(s_1(t) \otimes s_1^*(t))(s_1(t) \otimes s_1^*(t))^H\} \\
&\quad - E\{(s_1(t) \otimes s_1^*(t))\} E\{(s_1(t) \otimes s_1^*(t))^H\} \\
&\quad - E\{(s_1(t) s_1^H(t))\} \otimes E\{(s_1(t) s_1^H(t))^*\}.
\end{aligned}
$$

The diagonal matrix $\boldsymbol{\Phi}$ is diag $(e^{j\theta_1}, e^{j\theta_2}, ..., e^{j\theta_L})$ and diag $(e^{-j\pi\sin\theta_1}, e^{-j\pi\sin\theta_2}, ..., e^{-j\pi\sin\theta_L})$ in the mode-space transform and the array interpolation transform, respectively. $\boldsymbol{\Phi}^q$ is the product of q $\boldsymbol{\Phi}$s. Since D_q is a rank-one matrix, C_q is rank deflation and we average all C_q as

$$C = \sum_{q=1}^{Q} C_q = \tilde{A}_1(\boldsymbol{\theta}) \left(\sum_{q=1}^{Q} D_q \right) \tilde{A}_1^H(\boldsymbol{\theta}),$$

to recover the rank. The computation complexity of C is moderate at the cost that it has no characteristic of the virtual array aperture extension.

4 SIMULATIONS

Consider two coherent signals impinge from $\boldsymbol{\theta} = [20°, 35°]^T$ on a UCA with 16 sensors. The radius of the UCA is equal to the wavelength of the signals. The number of snapshots is set to 5000.

In the first test, the Signal-To-Noise Ratio (SNR) is set to 10 dB. The DOA estimation results of two DOA estimators based on two preprocessing methods are shown in Figure 1. Figure 1 shows that the estimation result based on the mode-space transform has a sharper peak than that based on

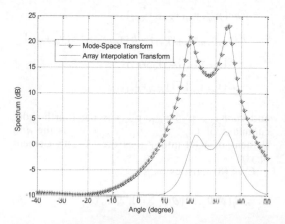

Figure 1.   The spatial spectrum of two DOA estimators.

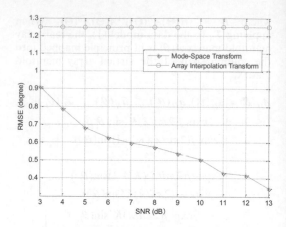

Figure 2.   The RMSE curves of two DOA estimators.

the array interpolation transform. In addition, Figure 1 shows that the estimator based on the mode-space transform has a little bias, while the estimator based on the array interpolation transform gives an obvious bias. The actual reason of the estimation bias phenomenon is not disclosed yet and needs to be studied in future. Monte-Carlo simulation is done in the second test, and SNR varies from 3 dB to 13 dB with the step 1 dB. The Root-Mean-Square Error (RMSE) curves of the DOA estimates by two estimators are shown in Figure 2. The statistical results in Figure 2 also indicate that the DOA estimates based on the mode-space transform outperforms those based on the array interpolation transform.

5 CONCLUSIONS

This paper has studied two preprocessing techniques, i.e., the mode-space transform and array interpolation transform, for the DOA estimation of coherent signals based on a uniform circular array in the fourth-order cumulant domain. Using two virtual ULA models given by the two preprocessing techniques, we have constructed a fourth-order cumulant matrix, together with the well-known spatial smoothing, to obtain two DOA estimators for coherent signals. The performance comparison of the two DOA estimators have been done by some theoretical analysis and simulations.

ACKNOWLEDGEMENTS

This work was sponsored by the fund project of the Science and Technology on Electronic Information Control Laboratory (201307310119-005-004).

The authors would like to thank Dr. Yun Hua and Shixin Yang for their help.

REFERENCES

B. Friedlander and A.J. Weiss, "Direction finding for wide-band signals using an interpolated array," Signal Processing, IEEE Transactions on, vol.41, no.4, Apr 1993, pp.1618–1634.

B. Porat and Benjamin Friedlander, "Direction finding algorithms based on high-order statistics," Signal Processing, IEEE Transactions on, vol.39, no.9, Sep 1991, pp.2016–2024.

M.C. Dogan and J.M. Mendel, "Applications of cumulants to array processing. I. Aperture extension and array calibration," IEEE Trans. on SP., vol.43, no.5, May 1995, pp.1200–1216.

P. Pal and P.P. Vaidyanathan, "Non uniform linear arrays for improved identifiability in cumulant based DOA Estimation," 2011 Conference Record of the Forty Fifth Asilomar Conference on, vol., no., Nov. 2011, pp.608–612, 6–9.

R.O. Schmidt, Multiple emitter location and signal parameter estimation, IEEE Trans. on AP., vol.34, no.3, Mar. 1986, pp.276–280.

Wang Yongliang, "The theory and algorithm of spatial spectrum estimation," Tsinghua University Press., Beijing, Dec. 2004, pp.344–346.

Electrical, Control Engineering and Computer Science – Liu (Ed.)
© 2016 Taylor & Francis Group, London, ISBN 978-1-138-02937-8

Adaptive low-voltage ride-through of DFIG based on crowbar with PDR and unloading protection circuit

Huilan Jiang, Zhe Jiang, Xing An & Jingpeng Wang
Smart Grid Key Laboratory of the Ministry of Education, Tianjin University, China

ABSTRACT: Crowbar circuit has become one of the most commonly used measures of the low-voltage ride-through technology. The traditional crowbar using fixed resistance is very difficult to suppress both rotor current and DC bus voltage and also cannot shorten the crowbar's input times and working time. Aiming at this problem, this paper analyzes the influence of crowbar and the DC side unloading protection circuit on rotor current and DC bus voltage when the Doubly Fed Induction Generator (DFIG) fails, and proposes an LVRT scheme of DFIG in which coordinates control the crowbar with a Parallel Dynamic Resistor (PDR) and unloading protection circuit. Combining to the transient process of DFIG, an adaptive control strategy for the scheme and the resistance setting method are formulated. The model of DFIG is built by Simulink, and the LVRT characteristics of the crowbar with PDR and unloading protection circuit are simulated during the voltage drop deeply, which are compared with the scheme of traditional crowbar with fixed resistance.

Keywords: crowbar; unloading protection circuit; adaptive control; Parallel Dynamic Resistor

1 INTRODUCTION

Environmental pollution and energy crisis have become a global problem; wind energy as a renewable clean energy finds favor with people, in which the wind turbines have great development.

A series of influence will be caused for the large-scale integration of wind power, and the capacity which wind turbine occupies in the power grid increases year by year. When the point voltage is dropped, if wind farms with large installed capacity are all off network, serious and even power system collapse may be caused, then the LVRT capability is required [1,2]. The method used is the most common one to improve the low-voltage ride-through capability of DFIG and increase the crowbar circuit in the rotor side or DC unloading circuit [3,4], but the suppression of rotor current and DC bus voltage cannot be both taken into account in the crowbar circuit of traditional or unloading circuit. Protection scheme of crowbar circuit and unloading circuit coordination has been proposed in the literature [5], to a certain extent, inhibition to the over current in the rotor and the bus over voltage are both considered, but when the voltage falling further, low voltage ride through capability needs to be improved.

Considering the traditional DFIG Crowbar circuit using a fixed resistance, it is difficult to give consideration to suppress the rotor current and the DC bus voltage and the control of the Crowbar input duration in the voltage drop significantly, a comprehensive scheme of low-voltage ride-through about DFIG, which, based on the parallel resistance of the Crowbar and the unloading circuit coordinated control, is proposed in this paper, and adaptive control strategy for parallel resistance and unloading circuit and the tuning method of the resistance are made with combination of DFIG transient process to realize the adaptive control of Crowbar parallel resistance and unloading circuit. By using the MATLAB/Simulink software platform to build model, the scheme for the simulation is analyzed.

2 RESISTANCE SETTING

2.1 Unloading protection circuit and crowbar with PDR

Since the increase in DC bus voltage lags the rotor current during the fault, the initial failure should first consider the problem of the rotor overcurrent. So the crowbar invested without PDR is to reduce the decay time constant of the rotor τ_r and control the overcurrent. When the DC bus voltage rises to its threshold value, the PDR and unloading protection circuit should be put to protect the capacitor. This method could suppress both the rotor current and DC bus voltage.

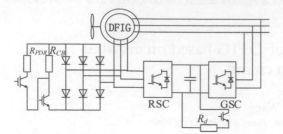

Figure 1. Crowbar with the PDR and unloading protection circuit.

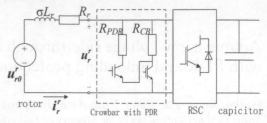

Figure 2. Rotor equivalent circuit.

2.2 Crowbar resistance and parallel dynamic resistance

The rotor equivalent circuit of considering the crowbar with parallel dynamic resistance is shown in Figure 2: During a fault, the rotor current is denoted by i_r^r (detailed derivation can be seen in reference [6]):

$$i_r^r = \frac{1}{\sigma L_r}\left[e^{-\frac{R_r}{\sigma L_r}t} \int e^{\frac{R_r}{\sigma L_r}t} u_r^r(t)dt - \frac{L_m}{L_s}\frac{sU_2 e^{j\omega_r t}}{R_r/\sigma L_r + j\omega_r} \right.$$

$$\left. + \frac{L_m}{L_s}(1-s)(U_1-U_2)\frac{e^{-(1/\tau_s + j\omega)t}}{R_r/\sigma L_r - 1/\tau_s - j\omega} \right]$$

(1)

The resistance setting of R_{CB} and R_{PDR} should be considered the most serious fault occurring when three-phase short circuit and the voltage drop depth is ($U_2 = 0$). Rotor fault current peak appeared in the post fault about T/2. In equation 1, with U_{dc_th}, T/2 and 0 instead of $u_r^r(t)$, t and U_2, we can obtain:

$$\left| i_{r\,\max}^r \right| = \left| \frac{U_{dc\_th}}{R_{requ}} + \frac{L_m}{\sigma L_r L_s}\frac{(1-s)U_1 e^{-(1/\tau_s + j\omega)T/2}}{R_{requ}/\sigma L_r - 1/\tau_s - j\omega} \right|$$

(2)

where the equivalent resistance of the rotor is $R_{requ} = R_r + R_{CB}$.
So the maximum value of rotor voltage is expressed as follows:

$$\left| u_{r\,\max}^r \right| = \left| i_{r\,\max}^r \right| R_{CB}$$

$$= \left| \frac{U_{dc\_th}}{R_{requ}} + \frac{L_m}{\sigma L_r L_s}\frac{(1-s)U_1 e^{-(1/\tau_s + j\omega)T/2}}{R_{requ}/\sigma L_r - 1/\tau_s - j\omega} \right| R_{CB}$$

(3)

In the protection scheme, since the main objective of investing crowbar at fault initial is to suppress the rotor current, the R_{CB} is tuning with the following principles: The rotor voltage $\left| u_{r\,\max}^r \right|$ is controlled within U_{dc_th}, and the R_{CB} could take a larger value. When the crowbar and PDR are both input, the equivalent parallel resistance R_{CB}' is as follows:

$$R_{CB}' = \frac{R_{CB}R_{PDR}}{R_{CB} + R_{PDR}}$$

(4)

If the crowbar and PDR are both input at t_1 time, it should focus on suppressing DC bus over voltage. In order to suppress the better, the rotor voltage maximum $\left| u_{r\,\max}^r \right|$ should be controlled within $0.2U_{dc\_th}$ when tuning the R_{PDR}:

$$\left| i_r^r(t_1) \right| R_{CB}' \leq 0.2U_{dc\_th}$$

(5)

The calculated:

$$R_{PDR} \leq \frac{0.2R_{CB}U_{dc\_th}}{R_{CB}\left| i_r^r(t_1) \right| - 0.2U_{dc\_th}}$$

(6)

2.3 DC unloading resistance

DC unloading circuit can effectively reduce the DC voltage fluctuations in the grid voltage drop, and improve DC capacitor life. For large-scale wind power systems, the DC capacitor of relatively large capacity, high cost makes it more meaningful to improve service life. By control of unloading circuit mainly by the input, the output of the active power and the DC side voltage, and through calculation and judgment, these data to determine whether the unloading circuit needs to be committed.
The control equation of DC bus voltage is:

$$\frac{1}{2}C\frac{dU_{dc}^2}{dt} = P_{in} - P_{out} = \Delta P.$$

(7)

where C is the capacitance, U_{dc} is the DC bus voltage, and P_{in} and P_{out} are the active power of DC input and output.

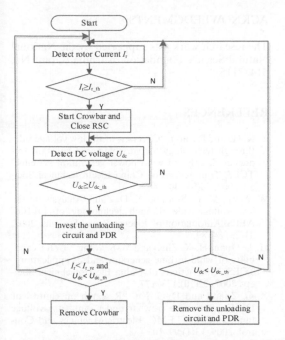

Figure 3. Adaptive control strategy.

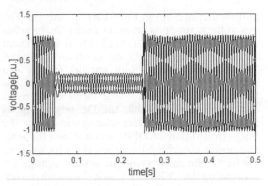

(a) Voltage drop

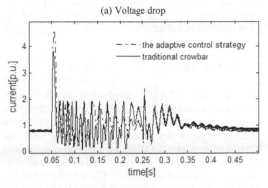

(b) Rotor current

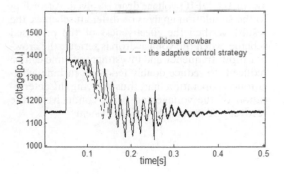

(c) DC bus voltage

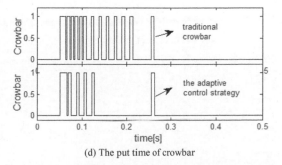

(d) The put time of crowbar

Figure 4. The simulation result of two schemes.

According to the energy conservation, the unloading resistance is as follows:

$$R_d = \frac{U_{dc\_th}^2}{\Delta P_{max}}. \tag{8}$$

3 ADAPTIVE CONTROL STRATEGY

Scheme of parallel resistance and unloading circuit coordinated control is as follows:

1. When the rotor current Ir exceeds its threshold, input the crowbar circuit, and lock the rotor side converter RSC;
2. When the DC bus voltage U_{dc} reaches its threshold, access parallel dynamic resistance R_{PDR}, and the unloading circuit; when U_{dc} falls below the threshold, disconnect R_{PDR} and unloading circuit;
3. When the U_{dc} is below threshold and rotor current Ir down to its return value, remove crowbar.

4 SIMULATION

The fault simulated is a serious three-phase grounding at the export bus of DFIG which begins at t = 0.05 s and clears at t = 0.25s, and the voltage dip is 0.8 p.u. The simulation result is shown in Figure 4, and two schemes are able to effectively inhibit the increase in rotor current, but the DC bus voltage suppression effect under the crowbar scheme of fixed resistors is obviously insufficient and leads to voltage limit and the recovery slower. The Crowbar scheme with parallel resistor can effectively balance the rotor current and DC bus voltage. It makes the crowbar input less number and invest less time while the DC bus voltage is lower than 1380V. So this scheme can reduce the uncontrolled time of DFIG and decrease the reactive power absorbing from the grid, which is conducive to the recovery of the system voltage. By comparing the LVRT index, as it shows, when voltage sag seriously, the effect of LVRT with parallel resistor crowbar and unloading circuit coordination control is better.

5 SUMMARY

The scheme of LVRT based on a parallel resistance and unloading circuit coordinated control is presented in this paper. Besides, the adaptive control strategy is developed by combining with the transient process of DFIG, and the tuning method of crowbar resistance, shunt resistance, and DC unloading resistance is given. The problem that the traditional crowbar cannot give consideration to the control of the rotor current and the DC bus voltage of the DFIG voltage drop is solved. According to the simulation analysis of different schemes, the results verified the effectiveness of the proposed scheme, when the voltage drop is serious, the crowbar input frequency and investment of time can be reduced, to reduce doubly fed wind turbine asynchronous operation time, thus delaying the deterioration of the voltage and being conducive to the recovery of the stability of the system.

ACKNOWLEDGMENTS

The research work was supported by the National Natural Science Foundation of China (Grant No. 51477115).

REFERENCES

Y.-K. He and P. Zhou, "Overview of the low voltage ride-through technology for variable speed constant frequency doubly fed wind power generation systems," TCES. Transactions of China Electrotechnical Society 2009, 24(9):140–146.

W. Wang, M.-D. Sun and X.-D. Zhu. "Analysis on the low voltage ride through technology of DFIG," AEPS. Automation of Electric Power Systems, 2007, 31(23):84–89.

H.-L. Guan, H.-X. Zhao and W.-S. Wang. "LVRT capability of wind turbine generator and its application," TCES. Transactions of China Electrotechnical Society, 2007, 22(10):173–177.

X.-G. Zhang and D.-G. Xu. "Research on control of DFIG with active crowbar under symmetry voltage fault condition," EMC. Electric Machines and Control, 2009, 13(1):99–103.

Y. Liu, X.-H. Hu and T. Chen. "Simulation Analysis and Experimental Study on Low Voltage Ride-Through Technology of Doubly-Fed Induction Generators," PSCE. Power System and Clean Energy, 2014, 30(4):47–52+59.

J. Yang, Dorrel D.G and Fletcher J.E. "A new converter protection scheme for doubly-fed induction generators during disturbances," Proceedings of the 34th Annual Conference of IEEE Industrial Electronics, November 10–13, 2008, Orlando, FL, USA:2100–2105.

Electrical, Control Engineering and Computer Science – Liu (Ed.)
© *2016 Taylor & Francis Group, London, ISBN 978-1-138-02937-8*

Research on the influence of load rate on district line loss rate in different areas of Jiangsu province

Guo-dong Gu
Jiangsu Electric Power Company, Nanjing, Jiangsu Province, China

ABSTRACT: Jiangsu Province can be divided into south, middle, and north three parts, the relationships between district load rate and district line loss rate of the three parts are different due to their different geographical locations, different levels of economic development and different power network structures. In order to investigate the difference, methods like classification and curve-fitting were used in this paper to get the relationship between district line loss rate and district load rate of Changzhou, Nanjing, and Xuzhou based on the marketing data provided by Jiangsu Electric Power Company. Similarities and differences were found by making a comparison among the three areas which can be instructive for the power company to realize fine management of regional power grid.

Keywords: district line loss rate; load rate; marketing data; comparison

1 INTRODUCTION

Power consumption information collection system is an important part of building the smart grid (Hu 2014). Some developed countries in Europe and America have already done some deep research in this field (Hart 2008, Farhangi 2010). This information collection system can effectively improve the marketing management level of power companies (Chen 2011). The marketing data got from the system include many parameters that can indicate the power demand as well as management level of a transformer district, namely the number of users (N), load rate (Rl), line loss rate (Dp), power supply volume (P), transformer capacity (Ct), and the average capacity of the users (dc).

The number of users (N) can reflect the size of the transformer district. District load rate (Rl) is the average load rate of a month which is defined as follows:

$$Rl = \frac{P/(24 \times 30)}{Ct} \qquad (1)$$

where P is the power supply volume of a month, Ct is the transformer capacity, 24 means the hours of one day and 30 means the days of a month.

District load rate (Rl) can reflect if the transformer was selected appropriate or not, in other words, Rl can reflect whether the installed capacity of the transformer matches the demand capacity of the district.

The district line loss rate (Dp) is defined as follows:

$$Dp = \frac{Ps}{P} \qquad (2)$$

where P is the power supply volume of a month, and Ps is the power actually consumed by users in a month.

Among the district parameters referred above, the load rate (Rl) and line loss rate (Dp) are dynamic parameters which are significant for daily management of district line loss. Therefore, in order to investigate the different relationships between district line loss rate and district load rate of southern, middle, and northern part of Jiangsu province, representative city of each region is selected. Different influences of load rate to line loss rate can be found out by making a comparison among the three cities based on their marketing data.

2 SAMPLES AND INSTRUCTIONS

Changzhou, Nanjing, and Xuzhou City were, respectively, selected as the typical cities of southern, middle, and northern regions of Jiangsu province. Firstly, marketing data of the three cities were pretreated to exclude the abnormal districts. After that, according to previous studies, the district average capacity of the users has the characteristic of concentration distribution in partitions.

Table 1. Divided range of the three cities.

City	Average capacity level (dc)	Urban power grid [kVA]	Rural power grid [KVA]
Nanjing	Low	3~7	3~6
	Middle	7~11	6~9
	High	11~14	–
Changzhou	Low	1~7	2~6
	Middle	7~11	6~10
	High	11~14	10~14
Xuzhou	Low	1~5	1~6
	Middle	7~11	6~9

Therefore, in order to eliminate the influence of average capacity to district line loss rate, urban districts as well as rural districts were divided into several types according to the concentration distribution situation of average capacity (Table 1), and in each type of districts the influences of load rate to line loss rate will be compared among the three cities.

According to Table 1, districts are generally divided into three types based on the size of average capacity. The district that has a lower average capacity belongs to the low dc type while that has a higher average capacity belongs to the high dc type and the others belong to the middle dc type; as for the division result, in urban power grid, both Nanjing and Changzhou can be divided into low, middle, and high average capacity three parts while Xuzhou can only be divided into low and middle two parts. In rural power grid, only Changzhou can be divided into low, middle, and high average capacity three parts while Nanjing and Xuzhou can only be divided into low and middle average capacity two parts. The average capacity reflects the consumption level of residents and it is concluded that the southern part of Jiangsu province has the highest electricity consumption level while the northern part of Jiangsu province has the lowest electricity consumption level.

3 INFLUENCE OF LOAD RATE ON LINE LOSS RATE

Figures 1–6 are the Dp-Rl fitting relationships in different dc levels of the three cities. Figures 1–3 are for the urban power grid and Figures 4–6 are for the rural power grid. Statistical line loss rate still includes fixed line loss rate and changing line loss rate though the data has been pretreated. And fixed line loss rate is inversely proportional to load rate while changing line loss rate is proportional to load rate. Therefore, the trend of the curve is

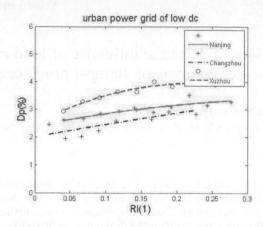

Figure 1. Dp-Rl fitting relationships in low dc level of the three cities in urban power grid.

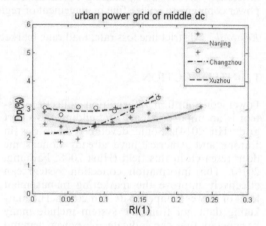

Figure 2. Dp-Rl fitting relationships in middle dc level of the three cities in urban power grid.

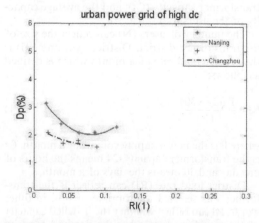

Figure 3. Dp-Rl fitting relationships in high dc level of the three cities in urban power grid.

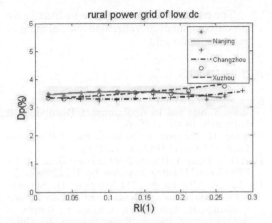

Figure 4. Dp-Rl fitting relationships in low dc level of the three cities in rural power grid.

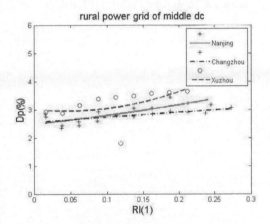

Figure 5. Dp-Rl fitting relationships in middle dc level of the three cities in rural power grid.

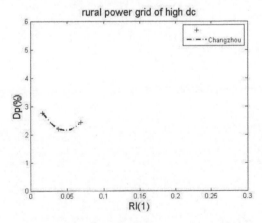

Figure 6. Dp-Rl fitting relationships in high dc level of the three cities in rural power grid.

related to the different proportions of fixed line loss rate and changing line loss rate (Sun 2012). As a general rule, district line loss rate has a downward trend followed by an upward trend with the increase in load rate.

In urban power grid, to low dc districts, the line loss rates of the three cities have an upward trend with the increase in load rate and the load rate has a larger influence on line loss rate in Xuzhou; To middle dc districts, the line loss rate of Changzhou increases significantly while the other two cities' line loss rate changes slightly. To high dc districts, the load rate is lower and the highest load rate of Changzhou is only 0.1. When the load rate is lower than 0.1, the line loss rate of Nanjing and Changzhouhasa downward trend for fixed line loss rate is the dominant factor. However, when the load rate is higher than 0.1, the dominant changes to changing line loss rate and the line loss rate of Nanjing have an upward trend.

In rural power grid, to low dc districts, the load rate has almost no influence on line loss rate; to middle dc districts, the line loss rates of the three cities have a slight upward trend with the increase in load rate and the load rate has a larger influence on line loss rate in Xuzhou and Nanjing; to high dc districts, only Changzhou has enough districts that belong to this type. The line loss rate of Changzhou has a downward trend when load rate is lower than 0.05 followed by an upward trend when load rate is higher than 0.05 with the increase in load rate.

4 CONCLUSIONS

In order to get the different relationships between district line loss rate and district load rate of southern, middle and northern part of Jiangsu province, Changzhou, Nanjing, and Xuzhou were selected as the typical cities, respectively. First, relationships are obtained by doing some classification and curve fitting based on the marketing data of the three cities. Then some regional characteristics of district line loss rate in southern, middle, and northern parts of Jiangsu province are inferred as follows:

As a general rule, the district line loss rate has a downward trend followed by an upward trend with the increase in load rate. This is mainly caused by different proportions of fixed line loss rate and changing line loss rate under different load rates.

1. For urban power grid:
   Among the low dc districts, Rl of northern part of Jiangsu has a larger impact on Dp while in southern and middle part the influence is smaller. To middle dc districts, Rl of southern part of Jiangsu has a larger impact on Dp, while

in the other two parts the influence is smaller. To high dc districts, the influence of Rl on Dp in the middle part of Jiangsu is larger than the southern part of Jiangsu.

2. For rural power grid:

To low dc districts, the load rate of the three parts of Jiangsu has almost no influence on line loss rate; To middle dc districts, the load rate has a larger influence on line loss rate in northern and middle part of Jiangsu; To high dc districts, comparison cannot be made for only southern part of Jiangsu has enough districts that belong to this kind.

Besides, district load rate of urban power grid has a larger impact on district line loss rate while in rural power grid the influence is smaller. This may be caused by different user type distributions as well as network constructors. More work need to be done to investigate the real reasons.

The conclusions above may be instructive for power companies to realize fine management of regional power grid.

REFERENCES

Chen Sheng, Lu Min, Power Usage Information Acquisition System and Its Application, J. Distribution & Utilization. 04 (2011) 45–49.

Farhangi H, The path of the smart grid, J. Power and Energy Magazine, IEEE. 8.1 (2010) 18–28.

Hart D.G., Using AMI to Realize the Smart Grid, J. IEEE PES General Meeting, July 2008. pp. 1–2 (2008):1–2.

Hu Jiang-yi, Zhu En-guo, Du Xin-gang, Du Shu-wei, Application Status and Development Trend of Power Consumption Information Collection System, J. Automation of Electric Power Systems. 02 (2014) 131–135.

Sun Hui, Low voltage distribution line loss factors and loss reduction measures, J. Public Communication of Science & Technology. 15 (2012) 103–104.

Electrical, Control Engineering and Computer Science – Liu (Ed.)
© *2016 Taylor & Francis Group, London, ISBN 978-1-138-02937-8*

Fault tolerant synchronization of chaotic system with sampled-data controller

Tao Ren, Yan jie Xu & Miao miao Liu
Software College, Northeastern University, Shenyang, China

Jian-xin Wen
North Automatic Control Technology Institute, Taiyuan, China

Sen He
Shenyang Jianzhu University, Shenyang, China

ABSTRACT: For the chaotic synchronization problem, a sampled-data synchronization controller is designed. The discontinuous error system was transformed to a discrete-time system by the Euler approximate discrete method. The faults in controller are considered also and based on the Lyapunov stability theory as well as the Linear Matrix Inequality (LMI) optimization technique, the sampled-data fault tolerant synchronization is proposed. Furthermore, the proposed synchronization control method is applied to the Chua's circuit and the validity is confirmed by numerical simulation.

Keywords: fault tolerant; synchronization control; sampled-data; chaotic system

1 INTRODUCTION

Chaotic system is a complex nonlinear system with many interesting features. Since 1990s, chaotic synchronization phenomena have been found in various fields, and received extensive attention of the researchers. Recently, various synchronization control methods have been reported [1–10].

Because the faults on sensors, actuators, and components can occur in chaotic system inevitably, it is necessary to design the fault tolerant synchronization controller for chaotic system. Very recently, several methods are proposed to increase the reliability of that kind of system [1–8]. In [1], a fuzzy controller is used to generate the fault tolerant control signal, and the proposed method is successfully applied to the chaotic model-tracking control between Lorenz system and Rossler system. In [2], a reliable feedback controller is established to guarantee synchronization between the master and slave chaotic systems even though some control component (actuator) faults occur. In [3], the impulsive robust fault-tolerant feedback control problem is discussed for chaotic Lur'e systems. The sufficient condition of uncertain Lur'e systems possessing integrity against actuator faults is given. In [4], the author proposed a fault-tolerant master-slave synchronization method by using time delay feedback control. Some new delay-dependent criteria are derived to satisfy the synchronization.

In [5], the fault tolerant synchronization of chaotic gyroscope systems versus external disturbances and faults is investigated and two techniques are considered as control methods. The robust fault-tolerant control problem against network faults and time-delays of a class of nonlinearly coupled chaotic systems are addressed in [6], an adaptive sliding mode control strategy is proposed. In [7], the fuzzy sampled-data controller, which contains a state feedback controller and a fault compensator is designed. In [8], an adaptive SMC approach to the robust synchronization control problem for a class of chaotic systems with actuator faults and saturation has been proposed.

Recently, with the development of high-speed computers and microelectronics, the sampled-data control systems attract widespread attentions which are used to produce a discrete-time control signal [11–13]. However, there is few work taking both the sampled-data control and the controller faults into account, which is practical significance and it is the motivation of the present study.

In this paper, for the chaotic synchronization system, the sampled-data controller is proposed. The controller faults are considered, then a fault tolerant synchronization method is investigated. The sufficient conditions are obtained to guarantee the synchronization of the master and slave systems based on LMI. Chua's circuit is chosen to illustrate the effectiveness of the proposed approach.

2 PROBLEM FORMULATION

Consider the following master and slave chaotic systems,

Master system: $\dot{x}_m(t) = Ax_m(t) + Bg(x_m(t))$ (1)

Slave system: $\dot{x}_s(t) = Ax_s(t) + Bg(x_s(t)) + u^F(t)$ (2)

where $g(\cdot)$ is a known nonlinear function; $u^F(t)$ is the fault controller which is described as follows

$$u^F(t) = Mu(t) \qquad (3)$$

where $M = diag\{m_1, m_2, ..., m_n\}$ is the actuator fault matrix with the following property, $0 \le m_i^{min} \le m_i \le m_i^{max} \le 1$, $i = 1, 2, ..., l$. When the ith actuator has no faults, $m_i^{min} = m_i^{max} = 1$; when the ith actuator is disabled, $m_i^{min} = m_i^{max} = 0$; when the ith actuator has partial faults, $0 < m_i < 1$. $u(t)$ is designed as the following sampled-data controller

$$u(t) = K(x_m(t_k) - x_s(t_k)), \quad t_k < t \le t_{k+1} \qquad (4)$$

where t_k is the sampling instant, define the following notations

$$M_0 = diag\{m_{01}, m_{02}, ..., m_{0n}\} \qquad (5)$$

$$H = diag\{h_1, h_2, ..., h_n\} \qquad (6)$$

$$G = diag\{g_1, g_2, ..., g_n\} \qquad (7)$$

$$|G| = diag\{|g_1|, |g_2|, ..., |g_n|\} \qquad (8)$$

where $m_{0i} = (m_i^{max} + m_i^{min})/2$, $g_i = (m_i - m_{0i})/m_{0i}$, $h_i = (m_i^{max} - m_i^{min})/(m_i^{max} + m_i^{min})$, According to (5)–(8), we can get

$$M = M_0(I + G), \quad |G| \le H \le I \qquad (9)$$

By Euler approximation discrete method, the discontinuous system can be transformed as

$$x_m(k+1) = (T_sA + I)x_m(k) + T_sBg(x_m(k)) \qquad (10)$$

$$x_s(k+1) = (T_sA + I)x_s(k) + T_sBg(x_s(k)) + T_su^F(k) \qquad (11)$$

$$u^F(k) = MK(x_m(k) - x_s(k))$$

where T_s is the sampling period of the controller. Defining the synchronization error as

$$\varrho(k) = x_m(k) - x_s(k) \qquad (12)$$

Then we get $u^F(k) = MKe(k)$, and the synchronization error system is described as

$$e(k+1) = (\bar{A} - \bar{K})e(k) + \bar{B}\eta(k) \qquad (13)$$

where $\eta(k) = g(x_m(k)) - g(x_s(k))$, $\bar{A} = T_sA + I$, $\bar{B} = T_sB$, $\bar{K} = T_sMK$.

Assumption 1: The nonlinear function $g(\cdot)$ of chaotic system (1) and (2) is satisfied in the following inequality at the interval $[0, \rho]$

$$0 \le \frac{\eta_i(d_i^Te, y)}{d_i^Te} = \frac{g_i(d_i^Te + d_i^Ty) - g_i(d_i^Ty)}{d_i^Te} \le \rho \qquad (14)$$

Lemma 1: For any appropriate dimension matrices X, Y, and a real symmetric positive definite matrix R, the following inequality holds:

$$XY + Y^TX^T \le XRX^T + Y^TR^{-1}Y \qquad (15)$$

3 MAIN RESULTS

Theorem 1 For given positive constant ε, the error system (13) is asymptotically stable by the reliable synchronization controller (12), if there exists $\Lambda = diag\{\lambda_1, \lambda_2, ..., \lambda_{n_h}\} \ge 0$, positive-definite matrices P and any appropriate dimension matrix Q such that the following inequality holds

$$\begin{bmatrix} -P & \rho D^T\Lambda & \bar{A}^TP - T_sQ & \varepsilon^{-1}T_sK^TH^T & 0 \\ & -2\Lambda & \bar{B}^TP & 0 & 0 \\ & & -P & 0 & \varepsilon PM_0 \\ * & & & -\varepsilon^{-1}I & 0 \\ & & & & -\varepsilon I \end{bmatrix} < 0 \qquad (16)$$

where $\bar{A} = T_sA + I$, $\bar{B} = T_sB$. The controller gain matrix can be calculated as $K = M_0^{-1}P^{-1}Q^T$.

Proof: Consider the following discrete Lyapunov function for the error system (13)

$$V(k) = e^T(k)Pe(k) \qquad (17)$$

where $P = P^T > 0$, then we can get

$$\Delta V(k) = e^T(k)[(\bar{A} - \bar{K})^T P(\bar{A} - \bar{K}) - P]e(k) + 2e^T(k)(\bar{A} - \bar{K})^T P\bar{B}\eta(k) + \eta^T(k)\bar{B}^T P\bar{B}\eta(k) \qquad (18)$$

According to Assumption 1, for any $\Lambda = diag\{\lambda_1, \lambda_2, ..., \lambda_{n_h}\} \ge 0$, we have

$$0 \le -2\eta^T\Lambda(\eta - \rho De) \qquad (19)$$

30

Combining (18) with (19), we obtain

$$\Delta V(k) \le \xi^{\mathrm{T}}(k)\Theta\xi(k) \qquad (20)$$

where

$$\Theta = \begin{bmatrix} (\bar{A}-\bar{K})^{\mathrm{T}}P(\bar{A}-\bar{K})-P & (\bar{A}-\bar{K})^{\mathrm{T}}P\bar{B}+\rho D^{\mathrm{T}}\Lambda \\ * & \bar{B}^{\mathrm{T}}P\bar{B}-2\Lambda \end{bmatrix},$$

$$\xi(k) = \begin{bmatrix} e^{\mathrm{T}}(k) & \eta^{\mathrm{T}}(k) \end{bmatrix}^{\mathrm{T}}.$$

Define Ψ

$$= \begin{bmatrix} -P & \rho D^{\mathrm{T}}\Lambda & \bar{A}^{\mathrm{T}}P-T_sQ & \varepsilon^{-1}T_sK^{\mathrm{T}}H^{\mathrm{T}} & 0 \\ & -2\Lambda & \bar{B}^{\mathrm{T}}P & 0 & 0 \\ & & -P & 0 & \varepsilon PM_0 \\ & * & & -\varepsilon^{-1}I & 0 \\ & & & & -\varepsilon I \end{bmatrix}$$
$$(21)$$

From Theorem 1, we have $\Psi < 0$. According to Schur Complement, we have

$$\Phi = \begin{bmatrix} \Phi_{11} & \rho D^{\mathrm{T}}\Lambda & \bar{A}^{\mathrm{T}}P-T_sQ \\ & -2\Lambda & \bar{B}^{\mathrm{T}}P \\ * & & \Phi_{33} \end{bmatrix} < 0 \qquad (22)$$

where $\Phi_{11} = -P + \varepsilon^{-1}T_s^2K^{\mathrm{T}}H^{\mathrm{T}}HK$, $\Phi_{33} = -P + \varepsilon PM_0M_0^{\mathrm{T}}P$. Use Lemma 1 and 2, the following inequality holds

$$E = \begin{bmatrix} -P & \rho D^{\mathrm{T}}\Lambda & E_{13} \\ & -2\Lambda & \bar{B}^{\mathrm{T}}P \\ * & & -P \end{bmatrix} < 0 \qquad (23)$$

where $E_{13} = \bar{A}^{\mathrm{T}}P - T_sQ - T_sK^{\mathrm{T}}G^{\mathrm{T}}M_0^{\mathrm{T}}P$. Let $Q = K^{\mathrm{T}}M_0^{\mathrm{T}}P$ and substitute $M = M_0(I+G)$ and $\bar{K} = T_sMK$ to (23), we can get the following inequality

$$E = \begin{bmatrix} -P & \rho D^{\mathrm{T}}\Lambda & (\bar{A}-\bar{K})^{\mathrm{T}}P \\ & -2\Lambda & \bar{B}^{\mathrm{T}}P \\ * & & -P \end{bmatrix} < 0 \qquad (24)$$

According to lemma 2, the following inequality holds,

$$\Theta = \begin{bmatrix} (\bar{A}-\bar{K})^{\mathrm{T}}P(\bar{A}-\bar{K})-P & (\bar{A}-\bar{K})^{\mathrm{T}}P\bar{B}+\rho D^{\mathrm{T}}\Lambda \\ * & \bar{B}^{\mathrm{T}}P\bar{B}-2\Lambda \end{bmatrix} < 0$$
$$(25)$$

Because Θ is negative, then we can get $\Delta V(k) < 0$, the error system (13) is asymptotically stable. Then we can get the conclusion that the master system and slave system can achieve synchronization.

4 NUMERICAL SIMULATION

In this section, the Chua's circuit is chosen to illustrate the effectiveness of the proposed sampled-data synchronization controller. First, we set parameter $\varepsilon = 1$ and $M = diag\{0.3,1,0\}$.

The error states are shown in Figure 1, the error system can become stable, i.e. the controller can synchronize the master and slave chaotic systems when faults occur in the controller. The outputs of the controller are shown in Figure 2.

Finally, we will show the effectiveness of the proposed controller when there are not any faults in the controller. From Figures 3 and 4, we can see that the chaotic systems are synchronous and the synchronization time is less (about 0.5s) than the case with faults.

In conclusion, the proposed fault-torrent controller can achieve the synchronization of chaotic

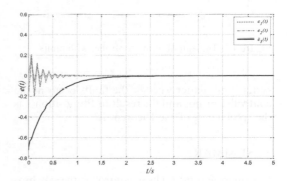

Figure 1. Curve of the error system without controller faults.

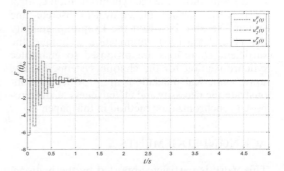

Figure 2. The output of the controller without faults.

31

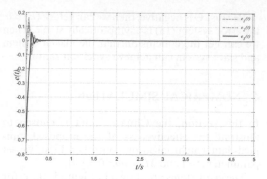

Figure 3. Curve of the error system with controller faults.

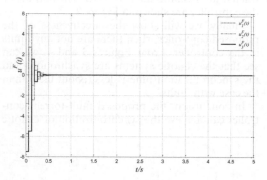

Figure 4. The output of the controller with faults.

systems no matter whether the faults occur in the controller or not and the result in Proposition 1 is the special case of the Theorem 1.

5 SUMMARY

Considering some faults in controller, the synchronization based on sampled-data control between the master and slave chaotic systems is investigated. First, by the Euler approximate discrete method, the discontinuous chaotic synchronization control plant is transformed into equivalent discrete-time system. Second, based on the Lyapunov stability theory and the LMI optimization technique, the sufficient conditions are proposed to guarantee synchronization for the chaotic systems. Finally, some numerical simulations are provided to demonstrate the necessity and effectiveness of our proposed synchronization method in this paper.

ACKNOWLEDGEMENTS

This work is partially supported by the National Natural Science Foundation of China (61473073, 61104074, 61203329), the Fundamental Research Funds for the Central Universities (N130417006, N110417005), and the Program for Liaoning Excellent Talents in University (LJQ2014028).

REFERENCES

[1] H.N. Wu and M.Z. Bai, "Active fault-tolerant fuzzy control design of nonlinear model tracking with application to chaotic systems," IET Control Theory and Applications, Vol.3, No.6, 2009, pp.42–653.

[2] H.H. Kuo, Y.Y. Hou and J.J. Yan, "Reliable synchronization of nonlinear chaotic systems," Mathematics and Computers Simulation, Vol.79, No.5, 2009, pp.1627–1635.

[3] Y. Zhang and J. Sun, "Impulsive robust fault-tolerant feedback control for chaotic Lur'e systems. Chao," Solitons and Fractals, Vol.39, No.3, 2009, pp.1440–1446.

[4] M.Y. Zhong and Q.L. Han, "Fault-tolerant master-slave synchronization for Lur'e systems using time-delay feedback control," IEEE Transactions on Circuits and Systems-I: Regular Papers, Vol.7, No.56, 2009, pp.1391–1404.

[5] F. Faezeh and A.S. Mahdi, "Fault tolerant synchronization of chaotic heavy symmetric gyroscope systems versus external disturbances via Lyapunov rule-based fuzzy control," ISA Transactions, Vol.51, No.1, 2012, pp.50–64.

[6] X.Z. Jin and G.H. Yang, "Adaptive sliding mode fault-tolerant control for nonlinearly chaotic systems against network faults and time-delays," Journal of the Franklin Institute, Vol. 350, No.5, 2013, pp.1206–1220.

[7] D.Z. Ma, H.G. Zhang, Z.S. Wang and J. Feng, "Fault tolerant synchronization of chaotic systems based on T-S fuzzy model with fuzzy sampled-data controller," China Physics B, Vol.5, No.19, 2010, pp.1–11.

[8] Y.H. Li and H.Y. Guang, "Fault tolerant control for a class of uncertain chaotic systems with actuator saturation," Nonlinear Dynamics, Vol.73, No.4, 2013, pp.2133–2147.

[9] T. Ren, Z.L. Zhu and H. Yu, "Design of finite-time synchronization controller and its application to security communication system," Applied Mathematics & Information Sciences, Vol.8, No.1, 2014, pp.387–391.

[10] T. Ren, Z.L. Zhu and H. Yu, M. Wang, "Sampled-data Synchronization Control of Chaotic Systems based on Min-Max Approach," Acta Physica Sinica, Vol.62, No.17, 2013, pp.1–8.

[11] A. Ichikawa and H. Katayama, "Linear time varying systems and sampled-data systems," Springer-Verlag Berlin, 26 February 2001.

[12] E. Fridam, U. Shaked and V. Suplin, "Input/output delay approach to robust sampled-data H-infinit control," Systems & Control Letters, Vol.54, No.3, 2007, pp.271–282.

[13] E. Fridman, A. Seuret and J. Richard, "Robust sampled-data stabilization of linear systems: an input delay approach," Automatic, Vol.40, No.8, 2004, pp.1441–1446.

Electrical, Control Engineering and Computer Science – Liu (Ed.)
© 2016 Taylor & Francis Group, London, ISBN 978-1-138-02937-8

Multi-interval schedule of economic dispatch for cogeneration systems

Ming-Tang Tsai & Fu-Sheng Chen
Cheng-Shiu University, Kaohsiung, Taiwan

Chi-Chun Lo
Chang Cung Memorial Hospital, Taiwan

ABSTRACT: The purpose of this paper is to solve a time-scheduling dispatch of cogeneration systems by considering the environmental protection. The fuel consumption and steam generation for a real cogeneration system will first be measured. Curve fitting method will then be used to get the I/O (Input/Output) curve of boilers, I/O cost curve of multi-fuels, and I/O cost curve of steam turbine. By using the relation between the fuels input and steam generation output, the emission models are derived in this paper. The objective function includes fuel cost, emission cost, and tie-line energy cost, subject to the use of mixed fuels, operational limits, and emission constraints. The Sequential Quadratic Programming (SQP) was used to solve the objective function by considering the Time-of-Use (TOU) and operational constraints. Results can provide a practical model for the cogeneration systems to solve the multi-interval schedule of economic dispatch problem.

Keywords: cogeneration systems; Time-of-Use; economic dispatch

1 INTRODUCTION

Cogeneration systems which are known as Combined Heat and Power (CHP) had now been extensively utilized by the industry. It offers a reliable, efficient, and economic mean to supply both thermal steam and electric power. The thermal steam and electric power generated from the cogeneration systems can be transmitted to buyer. Cogeneration systems bring some significant advantages due to the environmental aspects. It is a distributed energy sources, which can sell simultaneously the industrial thermal and electrical demand. It can also be constructed in urban areas and used as a distributed energy resources in the micro-grids [1–3]. In the last decades, consolidated cogeneration solutions had been developed and used in industrial applications [4]. Applications of cogeneration systems are still growing, more experience will be needed regarding the efficient operation for more energy saving.

To operate cogeneration systems more effectively, an efficient operational strategy has to be developed. At past, many papers had been published about cogeneration systems [5–12]. Cogeneration systems have to operate very efficiently to minimize its cost according the system schedule. Various fuels, such as Fuel Oil (F.O.),

Liquid Nature Gas (L.N.G.), and coal are available for dispatch at various cost bases. The optimal operating strategy determines the optimal distribution among the in-plant generation, fuels dispatch, and energy purchase to minimize the overall energy cost for a given electric and steam demand while satisfying the overall system constraints.

In this paper, the fuel consumption and steam generation will first be collected. Curve fitting method will then be used to get the Input-Output(I/O) curve for the heat input and the steam generation output. Multi-Fueled unit model was formulated to get the I/O curve for a unit burning mixed fuels. The emission model is presented as a function of fuel enthalpy. This paper will also consider the connection of the cogeneration system and utility company regarding the TOU rate [13]. The objective of optimal operation is to minimize the overall cost, which mainly includes fuel cost, emission cost, and tie-line power interchange cost while satisfying all constraints. The time-scheduling dispatch of cogeneration systems is a nonlinear problem. In this paper, Sequential Quadratic Programming (SQP) [14] was adopted to solve the optimal problem. Results can be shown that reasonable solutions provide a practical and flexible framework for cogeneration systems.

2 PROBLEM FORMULATIONS

The proposed algorithm was tested on a practical cogeneration. It contains 5 back-pressure steam turbine, 1 extraction condenser steam turbine, and 6 steam boilers. The fuels including F.O., L.N.G., and coal were used in each boiler for producing high-pressure steam. It is required to assess the economic and operational benefits through optimal steam loading allocation among boilers and turbines. Models determination for optimal operation evaluation as follows.

2.1 I/O cost curve of boilers

It is assumed that the I/O curve of a boiler is a 3-rd order polynomial, we have

$$F_{bi}(M_{bi}) = A_o + A_1 \times M_{bi} + A_2 \times M_{bi}^2 + A_3 \times M_{bi}^3$$
$$i = 1, 2, .. 6 \quad (1)$$

where

$F_{bi}(M_{bi})$: Enthalpy of the i-th boiler at i-th bus (MBTU/H).
M_{bi}: Steam output of the i-th boiler at i-th bus(T/H).
A_o, A_1, A_2, A_3: Coefficients of the I/O operation curve.

With mixed fuel used, a proper modification will be needed. The dual-fueled unit model was formulated in [15]. Equation (2) is used to represent a unit burning three fuels simultaneously. We have

$$F_{bT}(M_b(t)) = F_{b1}(M_b(t)) \times (\lambda_1(t) + \eta_{1/2} \times \lambda_2(t)$$
$$+ \eta_{1/3} \times \lambda_3(t)) \quad (2)$$

where

$F_{bT}(M_b(t))$: Total enthalpy at t-th interval (MBTU/H).
$F_{b1}(M_b(t))$: Enthalpy of fuel 1 at t-th interval (MBTU/H).
$\eta_{1/2}$: The efficiency ratio of fuel 1/fuel 2.
$\eta_{1/3}$: The efficiency ratio of fuel 1/fuel 3.
$\lambda_1(t), \lambda_2(t), \lambda_3(t)$: The mixed ratio of fuel 1,2, and 3 at t-th interval.
With $\lambda_1(t)+, \lambda_2(t)+ \lambda_3(t) = 1$

The I/O cost curve of boilers can be described by

$$FBC_T(t) = F_{bT}(M_b(t)) \times BC_T(t) \quad (3)$$

$$BC_T(t) = BC_1 \times \lambda_1(t) + BC_2 \times \lambda_2(t) + BC_3 \times \lambda_3(t) \quad (4)$$

where

$FBC_T(t)$: The total operation cost of the boilers at t-th interval (NT$)

$BC_T(t)$: The total cost of fuel mixture at t-th interval(NT$/MBTU)
BC_1, BC_2, BC_3: The cost of fuel 1, 2, and 3 (NT$/MBTU).

2.2 I/O operation curve of steam turbines

For back-pressure turbine generator, the power equation for turbine i can be formulated by [16]

$$P_{gi} = K_{oi} + K_{1i}M_{mi} + K_{2i}M_{mi}^2 \quad i = 1, 2, ..., 5 \quad (5)$$

For extraction-condense turbine generator, the power equation for turbine i can be written by [16]

$$P_{gi} = K_{oi} + K_{1i}M_{mi} + K_{2i}M_{wi} \quad i = 6 \quad (6)$$

where M_{mi} is the medium pressure extraction flow of i turbine. P_{gi} is the generated electric power from turbine i; K_{oi}, K_{1i}, K_{2i} are coefficients of turbine i which can then be found by curve fitting technique with field data.

2.3 Emission model

Two primary emissions are sulfur dioxide (SO_x) and nitrogen oxides (NO_x). Emission models may be defined as the amount of fuel consumed or as a function of boiler steam. The emission model with mixed fuels can be formulated by

$$E_{Si}(\bullet) = (\gamma_{S1i}\lambda_{1i}(t) + \gamma_{S2i}\lambda_{2i}(t) + \gamma_{S3i}\lambda_{3i}(t))$$
$$\times : F_{bTi}(M_{bi}(t), \lambda_{1i}(t), \lambda_{2i}(t), \lambda_{3i}(t)) \quad (7)$$

$$E_{Ni}(\bullet) = (\gamma_{N1i}\lambda_{1i}(t) + \gamma_{N2i}\lambda_{2i}(t) + \gamma_{N3i}\lambda_{3i}(t))$$
$$\times F_{bTi}(M_{bi}(t), \lambda_{1i}(t), \lambda_{2i}(t), \lambda_{3i}(t)) \quad (8)$$

$E_{Si}(\bullet)$: The amount of pollutant SO_x for the i-th boiler at the t-th interval (T/h).
$E_{Ni}(\bullet)$: The amount of pollutant NO_x for the i-th boiler at the t-th interval (T/h).
$F_{bTi}(\bullet)$: Total enthalpy for the i-th boiler at the t-th interval (T/h).
$\gamma_{S1i}, \gamma_{N1i}$: The emission factor of SO_x and NO_x with oil for the i-th boiler (T/Mbtu).
$\gamma_{S2i}, \gamma_{N2i}$: The emission factor of SO_x and NO_x with LNG for the i-th boiler (T/Mbtu).
$\gamma_{S3i}, \gamma_{N3i}$: The emission factor of SO_x and NO_x with coal for the i-th boiler (T/Mbtu).
$\lambda_{1i}(t), \lambda_{2i}(t), \lambda_{3i}(t)$: The mixed ratio of fuel 1, 2, and 3 for the i-th boiler at t-th interval.

2.4 The model of economic dispatch

The model of economic dispatch needs to meet the steam demand of in-plant process and electricity.

Table 1. The Time-of-Use rates (TOU Rate).

	Electricity sale price (NT$/kwh)		Utility Buy-back Price NT$/kwh
	Level-1	Level-2	
Peak period	3.04	2.7480	3.04
Semi-peak period	1.83	1.5767	1.83
Off-peak period	0.69	0.4729	0.69

Level-1: power exported under 20% rated capacity.
Level-2: power exported over 20% rated capacity.

The objective function included fuel cost, emissions cost, and interchange cost in the multi-interval schedule as shown follows.

$Min\,C()$

$$= \sum_{t=1}^{T} \left\{ \begin{array}{l} \sum_{i=1}^{6} BC_i(\lambda_{1i}(t), \lambda_{2i}(t), \lambda_{3i}(t)) \times F_{bi}(M_{bi}(t), \\ \lambda_{1i}(t), \lambda_{2i}(t), \lambda_{3i}(t)) \pm EC(t) \times P_{tie}(t) \\ + C_S \times \sum_{i=1}^{6} E_{Si}(M_{bi}(t), \lambda_{1i}(t), \lambda_{2i}(t), \lambda_{3i}(t)) \\ + C_N \times \sum_{i=1}^{6} E_{Ni}(M_{bi}(t), \lambda_{1i}(t), \lambda_{2i}(t), \lambda_{3i}(t)) \end{array} \right\} \cdot \tau$$

(9)

C_S and C_N are the charged pollution emission fee for SO_x and NO_x. $EC(t)$ is the TOU rates [13] as shown in Table 1. $P_{tie}(t)$ is the electricity purchased from or sold to the utility and τ is the time interval. The solutions of the objective function must meet the operational constraints. The constraints have equality constraints (power balance, steam balance for boilers, turbine, and industrial process) and inequality constraints (boilers, steam turbine, power generation, and emission limit).

3 SIMULATION RESULTS

The proposed algorithm was tested on a practical cogeneration. It contains 5 back-pressure steam turbine, 1 extraction condenser steam turbine, and 6 steam boilers. The emission limits for SOx and NOx are 0.9T/H and 0.5T/H, respectively. In this paper, the Sequential Quadratic Programming (SQP) [14] was adopted to solve the optimal problem. SQP solve a sequence of optimization problems. Each problem is optimized a quadratic model of the objective subject to the linear constraints. SQP is an efficient and powerful algorithm to appropriately solve this nonlinear problem. The detail formulations for SQP is described in [14]. In our study case, there are 44 variables, 13 equality constraints, and 5 inequality constraints. All programs were written with Matlab Toolbox Software. An IBM PC with Intel i3–3240 CPU and 4 DRAM is used in our test. The CPU time in the 24-hr scheduling dispatch is about 12.3 sec. with emission constraints and 10.1 sec. without emission constraints, respectively.

Figure 1 shows the relationships of generations, load, and purchased power in the 24 hr dispatch. From the Figure 1, due to the price of purchased power in the peak interval is higher than the generating cost, the output of cogeneration systems thus increases from 63.7 MW to 92.95 MW and the power is brought from 65.21 MW to 31.55 MW when the semi-peak interval is changed to the peak interval. Similarly, due to the price of power in the off-peak interval is cheaper than the generating cost, the output of cogeneration systems is reduced from 94.19 MW to 62.8MW and the power is brought from 28.42 MW to 60.10 MW.

Figure 2 shows the optimal generations of generators in the 24-hours schedule. Because Pg1~Pg5 are the back-pressure steam turbines, they are the lack of flexibility in the operational schedule. Pg6 is the extraction condenser steam turbine, which adjust power generations to achieve the optimal dispatch of the cogeneration systems.

Figure 3 shows the operational cost of the 24 hr scheduling dispatch. Due to the cost of purchasing power during the peak interval is more than the cost of purchasing power during semi-peak interval/off-peak interval, its operating cost will be greater. From the Figure 3, the operating cost in a daily dispatch is about 4520240 NT$ by considering

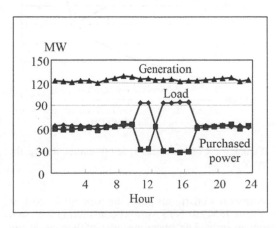

Figure 1. The relationships of generations, load, and purchased power.

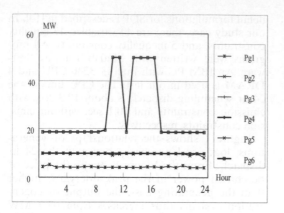

Figure 2. The optimal generations of generators in the 24-hours schedule.

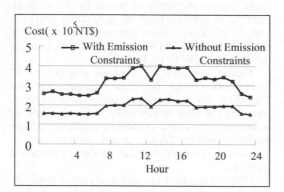

Figure 3. The optimal operation cost of the 24 hr scheduling dispatch.

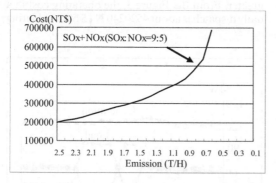

Figure 4. Cost-emission trade-off curves during the peak period.

emission constraints and the operating cost is about 7654720 NT$ without considering emission constraints. The operating cost with considering emission constraints is larger than the operating cost without considering emission constraints.

Figure 4 shows the relationship between emissions and operation cost. The emissions and operation cost are often conflict. Figure 4 shows cost-emission trade-off curves during the peak period. It provides the utility planners a wider range of alternatives showing the effects of various pollutants. Instead of using maximal allowable limits for emission as constraints, an appropriate strategy can be chosen to meet the desired level of emission and cost. If the environmental requirement is the higher, the relative amount of emission will be reduced and operating cost is relative increased.

4 CONCLUSIONS

This paper presents an optimal approach to solve a multi-scheduling dispatch of cogeneration systems by considering the environmental protection. By considering the Time-of-Use(TOU) between cogeneration systems and utility companies, the Sequential Quadratic Programming (SQP) was used to solve this problem. The optimization can handle the trade-off's between cost and emission with multi-fuel dispatch. The simulation also shows that the TOU rate significantly affects the overall operational cost. It also provides a practical and flexible framework for evaluating the emission control.

REFERENCES

[1] A.H. Azit, K.M. Nor, Optimal sizing for a gas-fired grid-connected cogeneration system planning. IEEE Transaction on Energy Conversion, 24, 2009, pp. 950–958.
[2] M. Motevasel, A.R. Seifi, T. Niknam, Multi-objective energy management of CHP (combined heat and power)-based micro-grid. Energy 51, 2013. pp. 123–136.
[3] A.K. Basu, A. Bhattacharya, S. Chowdhury, Planned scheduling for economic power sharing in a CHP-based micro-grid. IEEE transactions on Power Systems 27, 2012, pp. 30–38.
[4] F. Salgado, P. Pedrero, Short-term. Short-term operating planning on cogeneration system: A survey. Electric Power Systems Research 78, 2008, pp. 835–848.
[5] S. Daniel, C. Lou, K. Min, Economic models for cogeneration facilities and host utilities under the right to sell provision. Electric Power Systems Research, 103, 2013, pp. 214–222.
[6] U. Çakir, K. Çomakli, F. Yükse, The role of cogeneration systems in sustainability of energy. Energy Conversion and Management, 63, 2012, pp. 196–202.
[7] M.T. Tsay, W.M. Lin, J.L. Lee, Application of evolutionary programming for economic dispatch of cogeneration systems under emission constraints, International Journal of Electrical Power & Energy Systems, 23, 2001, pp. 805–812.

[8] L.L. Lai, J.T. Ma, J.B. Lee, Multi-time interval scheduling for daily operation of a two-cogeneration system with evolutionary programming. International Journal of Electrical Power & Energy Systems, 20, 1998, pp. 305–311.

[9] B.K. Chen, C.C. Hong, Optimum operation for a back-pressure cogeneration system under time-of-use rates. IEEE Transactions on Power Systems 11, 1992, pp. 1074–1084.

[10] S.A. Farghal, R.M. El-dewieny, A.M. Riad, Optimum operation of cogeneration plants with energy purchase facilities. IEE Proceedings Generation, Transmission, and Distribution, 134, 1987, pp. 313–319.

[11] O. Linkevičs, A. Sauhats, "Formulation of the Objective Function for Economic Dispatch Optimisation of Steam Cycle CHP Plants," 2005 IEEE Russia Power Tech, 2005, pp. 1–6.

[12] A. Dolgicers, S. Guseva, A. Sauhats, O. Linkevičs, A. Mahņitko, I. Zicmane, "Market and Environmental Dispatch of Combined Cycle CHP Plant,"IEEE Bucharest Power Tech Conference, 2009, pp. 1–6.

[13] Taiwan Power Company (TPC). Time-of-Use rate for cogeneration plants. The electricity rates structure for Taipower Company, April 2011.

[14] Matlab Optimization Toolbook User's Guide, The Math Works, Inc. 2011.

[15] Shoults R.R, Robert K.G. Power system operation-course lecture notes. Energy Systems Research Center, The University of Texas at Arlington, Arlington, Texas 76019, 1986.

[16] Lee J.L. Application of evolutionary programming to optimal dispatch of cogeneration system. The Master Thesis of National Sun Yat-Sen University, June 1999.

Electrical, Control Engineering and Computer Science – Liu (Ed.)
© 2016 Taylor & Francis Group, London, ISBN 978-1-138-02937-8

Fault location algorithm based on fault region determination for partial coupling four-circuit transmission lines

Weiming Luo, Huiqiang Wang, Jianhua Wang, Lin Zeng & Shiyi Chen
Dongguan Power Supply Bureau, Guangdong Power Grid Corporation, Dongguan, Guangdong Province, China

Ling Xiao & Wanlin Du
Guangzhou Jiayuan Electric Engineering Limited Company, Guangzhou, China

ABSTRACT: Partial coupling multi-parallel transmission lines on the same tower are widely used in power grid, but current location methods cannot achieve accurate fault location due to the structural complexity of partial coupling multi-parallel transmission lines. In this paper, according to the decoupling method of multi-parallel transmission lines, the voltage and current interface equations of coupling demarcation point are constructed to connect with the regions of lines. Then, based on the interface equations, fault analysis is performed to find out the characteristics for fault region determination of the partial coupling four-circuit transmission lines. Further, fault region determination criterion is proposed, and fault location equations of each fault region are constructed respectively to present a fault location algorithm of partial coupling four-circuit transmission lines. Finally, the simulation model for transmission lines is constructed with ATP-EMTP. As demonstrated by the analysis results, the proposed scheme can achieve accurate fault location in partial coupling multi-parallel transmission lines.

Keywords: multi-parallel transmission lines; fault location; fault region determination

1 INTRODUCTION

Due to the demand of transmission capacity and the lack of channel resources, multi-parallel transmission lines on the same tower are widely used in power grid to increase the transmission capacity per line unit and reduce the cost of electric power construction. In complex power grid, fault location is very important to repair fault lines, recover power supply and ensure the safe, stable and economical operation of power system. Currently, fault location methods mainly include travelling wave [1]–[3] and fault analysis, of which fault analysis can be further divided into frequency domain method [4], [5] and time domain method [6], [7]. Frequency domain method bases on fault power frequency electrical quantities and long line equation. It requires a long data time window, but has high accuracy and stability. Time domain method bases on the Bergeron model and electric instant quantities. Compared with frequency domain method, it requires a shorter data time window and a higher sampling frequency. Besides, its accuracy is easily influenced by the interference of high frequency components.

With the spread of multi-parallel transmission lines, there appear many fault location methods using decoupling algorithms, such as six-sequence component method and fault location frequency domain method of double-circuit transmission lines [8]–[10], twelve-sequence component method and fault location frequency domain method of four-circuit transmission lines [11], [12], and fault location method of multi-parallel transmission lines based on single-terminal electric quantities [13]. These methods can offer plenty of practicable schemes, but only apply to multi-circuit transmission lines completely on the same tower. However, in most cases, four-circuit transmission lines partially share a tower, while the rest are double-circuit or single-circuit. That is to say, partial coupling four-circuit transmission method connects with three, four or more terminals of power grid. Consequently, the fault analysis and fault location in multi-parallel transmission lines are very complex.

As proposed in Refs. [14]–[16], in fault branch determination and fault location schemes for three different double-circuit T connecting lines, each circuit line directly connects at t node, namely the line voltage of circuit line is equal to each other at t node. In contrast, for partial coupling multi-parallel transmission lines, electromagnetic coupling occurs at the demarcation point, which can result in unequal line voltage. Because of the great structural difference, the fault location schemes of T connecting lines are generally not suitable for partial coupling multi-parallel transmission lines. In view of this, this paper proposes a fault analysis and fault location frequency domain algorithm for partial coupling four-circuit transmission lines connecting three-terminal system. Firstly, based on the decoupling method of multi-parallel transmission lines, the voltage and current interface equations of coupling demarcation point are constructed to connect with the regions of lines. Then, based on the interface equations and using three-terminal fault voltage and current, their equal directional and circulating components are extracted, and the voltage of demarcation point is obtained with long line equation. Further, through analyzing the characteristics for fault region determination of the partial coupling four-circuit transmission lines, the fault region determination criterion is presented, and the fault location equations of each fault region are constructed respectively to present a fault location algorithm of partial coupling four-circuit transmission lines based on three-terminal electrical quantities. Finally, the simulation model for transmission lines is constructed with ATP-EMTP.

2 DECOUPLING METHODS FOR MULTI-PARALLEL TRANSMISSION LINES

This paper adopts the concept of decoupling methods proposed in Ref. [11]. For the four-circuit transmission lines with square shape, with un-transposition between the lines and with uniform transposition between the lines, we first make three-phase conductor equal to a single one and then analyze it, namely we regard 12 lines as four-parallel system. Thus, we can obtain the decoupling transform equation of four-circuit transmission lines based on the sagami transformation matrix of four-parallel system:

$$\begin{bmatrix} \dot{Y}^e \\ \dot{Y}^f \\ \dot{Y}^g \\ \dot{Y}^h \end{bmatrix} = \frac{1}{4} \begin{bmatrix} 1 & 1 & 1 & 1 \\ 1 & -1 & 1 & -1 \\ 1 & 1 & -1 & -1 \\ 1 & -1 & -1 & 1 \end{bmatrix} \begin{bmatrix} \dot{Y}^I \\ \dot{Y}^{II} \\ \dot{Y}^{III} \\ \dot{Y}^{IV} \end{bmatrix} \quad (1)$$

where $\dot{Y}^I$, $\dot{Y}^{II}$, $\dot{Y}^{III}$ and $\dot{Y}^{IV}$ respectively denote three-phase electrical quantitises (voltage or current) of I, II, III and IV-circuit lines. Eq. (1) divides four-circuit coupling electrical quantitises into 4 separate modulus components, i.e., $\dot{Y}^e$, $\dot{Y}^f$, $\dot{Y}^g$ and $\dot{Y}^h$, of which $\dot{Y}^e$ is the equal directional component with four-line overlap, and $\dot{Y}^f$, $\dot{Y}^g$ and $\dot{Y}^h$ respectively denote circulating f, g and h

components with four-line loop mutually. The single-line symmetrical components transform decouple phase-to-phase among coupling modulus three-phase lines. Thus, 12 separate modulus components can be obtained.

Similarly, with regard to decoupling methods for double-circuit transmission lines, we make three-phase conductor equal to a single one and then analyze it. Namely, we regard six lines as double-parallel system, and then obtain the decoupling transform equation of double-circuit transmission lines based on the sagami transformation matrix of double-parallel system:

$$\begin{bmatrix} \dot{Y}^e \\ \dot{Y}^f \end{bmatrix} = \frac{1}{2} \begin{bmatrix} 1 & 1 \\ 1 & -1 \end{bmatrix} \begin{bmatrix} \dot{Y}^I \\ \dot{Y}^{II} \end{bmatrix} \quad (2)$$

The electrical quantities can be divided into two separate modulus components. One is an equal directional component, and the other is a circulating one. Then, based on the symmetrical component method, modulus components can be decoupled among three phases to achieve six separate modulus components.

3 MODULUS COMPONENT INTERFACE MODEL OF COUPLING DEMARCATION POINT IN PARTIAL COUPLING FOUR-CIRCUIT TRANSMISSION LINES

Figure 1 shows a model of three terminal partial coupling four-circuit transmission lines, in which the transmission lines connects with a three-terminal system. M, J and K are the terminals. Point D is the coupling demarcation point. Line DM is the four-circuit coupling region and its length is L_{DM} (positive-sequence wave impedance and propagation parameter are respectively Z_{DM}^1 and γ_{DM}^1). Line DJ and DK are double-circuit coupling regions and their lengths are respectively L_{DJ} and L_{DK} (positive-sequence wave impedance and propagation parameter for DJ are respectively

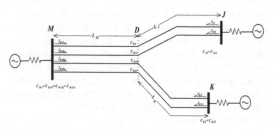

Figure 1. Three terminal partial coupling four parallel transmission lines.

Z^1_{DJ} and γ^1_{DJ}, and those for DK are respectively Z^1_{DK} and γ^1_{DK}).

When a fault occurs in partial coupling four-circuit transmission lines, it may locate in different coupling regions. Accordingly, the corresponding fault analysis grid is different. Fault analysis depends on coupling region. This paper only uses the positive-sequence component of each modulus component during fault analysis, fault region determination and fault location. The equal directional and circulating components of voltage and current respectively denote equal directional and circulating positive-sequence components. The positive direction of current quantities is defined as the direction from nodes to lines.

Decoupling the four-circuit and double-circuit transmission lines by Eqs. (1) and (2), each modulus voltage of terminals M, J and K can be obtained as: $\dot{U}^{el}_M, \dot{U}^{fl}_M, \dot{U}^{gl}_M, \dot{U}^{hl}_M, \dot{U}^{el}_J, \dot{U}^{fl}_J, \dot{U}^{el}_K, \dot{U}^{fl}_K$. The modulus current flowing into line from each terminal can be set as: $\dot{I}^{el}_M, \dot{I}^{fl}_M, \dot{I}^{gl}_M, \dot{I}^{hl}_M, \dot{I}^{el}_J, \dot{I}^{fl}_J, \dot{I}^{el}_K, \dot{I}^{fl}_K$. Each modulus voltage of coupling demarcation point D is different for different analysis regions. In Line DM: $\dot{U}^{el}_{D(DM)}, \dot{U}^{fl}_{D(DM)}, \dot{U}^{gl}_{D(DM)}, \dot{U}^{hl}_{D(DM)}$; In Line DJ: $\dot{U}^{el}_{D(DJ)}, \dot{U}^{fl}_{D(DJ)}$; In Line DK: $\dot{U}^{el}_{D(DK)}, \dot{U}^{fl}_{D(DK)}$. Analogously, each modulus current flowing into line from point D: $\dot{I}^{el}_{D(DM)}, \dot{I}^{fl}_{D(DM)}, \dot{I}^{gl}_{D(DM)}, \dot{I}^{hl}_{D(DM)}, \dot{I}^{el}_{D(DJ)}, \dot{I}^{fl}_{D(DJ)}, \dot{I}^{el}_{D(DK)}, \dot{I}^{fl}_{D(DK)}$.

According to the long line equation of transmission lines, the modulus voltage of point D, Line DM and modulus current flowing into line DM from point D can be deduced from each modulus voltage and current of terminal M:

$$\begin{cases} \dot{U}^{ml}_{D(M)} = \dot{U}^{ml}_M ch(\gamma^1_{DM}L_{DM}) - \dot{I}^{ml}_M Z^1_{DM} sh(\gamma^1_{DM}L_{DM}) \\ \dot{I}^{ml}_{D(M)} = \dfrac{\dot{U}^{ml}_M}{Z^1_{DM}} sh(\gamma^1_{DM}L_{DM}) - \dot{I}^{ml}_M ch(\gamma^1_{DM}L_{DM}) \\ \qquad\qquad (m = e, f, g, h) \end{cases}$$
(3)

where, only when there is no fault in line DM, $\dot{U}^{ml}_{D(M)} = \dot{U}^{ml}_{D(DM)}$ and $\dot{I}^{ml}_{D(M)} = \dot{I}^{ml}_{D(DM)}$.

Similarly, the modulus voltage of point D, Line DJ and modulus current flowing into line DJ from point J can be deduced from each modulus voltage and current of terminal J:

$$\begin{cases} \dot{U}^{ml}_{D(J)} = \dot{U}^{ml}_J ch(\gamma^1_{DJ}L_{DJ}) - \dot{I}^{ml}_J Z^1_{DJ} sh(\gamma^1_{DJ}L_{DJ}) \\ \dot{I}^{ml}_{D(J)} = \dfrac{\dot{U}^{ml}_J}{Z^1_{DJ}} sh(\gamma^1_{DJ}L_{DJ}) - \dot{I}^{ml}_J ch(\gamma^1_{DJ}L_{DJ}) \\ \qquad\qquad (m = e, f) \end{cases}$$
(4)

where, only when there is no fault in line DJ, $\dot{U}^{ml}_{D(J)} = \dot{U}^{ml}_{D(DJ)}$ and $\dot{I}^{ml}_{D(J)} = \dot{I}^{ml}_{D(DJ)}$.

The modulus voltage of point D, Line DK and modulus current flowing into line DK from point D can be deduced from each modulus voltage and current of terminal K:

$$\begin{cases} \dot{U}^{ml}_{D(K)} = \dot{U}^{ml}_K ch(\gamma^1_{DK}L_{DK}) - \dot{I}^{ml}_K Z^1_{DK} sh(\gamma^1_{DK}L_{DK}) \\ \dot{I}^{ml}_{D(K)} = \dfrac{\dot{U}^{ml}_K}{Z^1_{DK}} sh(\gamma^1_{DK}L_{DK}) - \dot{I}^{ml}_K ch(\gamma^1_{DK}L_{DK}) \\ \qquad\qquad (m = e, f) \end{cases}$$
(5)

where, only when there is no fault in line DK, $\dot{U}^{ml}_{D(K)} = \dot{U}^{ml}_{D(DK)}$ and $\dot{I}^{ml}_{D(K)} = \dot{I}^{ml}_{D(DK)}$.

From Eqs. (1) and (2), the modulus voltage and current interface equations of point D, Line DM are constructed by Lines DJ and DK:

$$\begin{cases} \dot{U}^{el}_{D(DM)} = 0.5\left(\dot{U}^{el}_{D(DJ)} + \dot{U}^{el}_{D(DK)}\right) \\ \dot{U}^{fl}_{D(DM)} = 0.5\left(\dot{U}^{fl}_{D(DJ)} + \dot{U}^{fl}_{D(DK)}\right) \\ \dot{U}^{gl}_{D(DM)} = 0.5\left(\dot{U}^{el}_{D(DJ)} - \dot{U}^{el}_{D(DK)}\right) \\ \dot{U}^{hl}_{D(DM)} = 0.5\left(\dot{U}^{fl}_{D(DJ)} - \dot{U}^{fl}_{D(DK)}\right) \end{cases}$$
(6)

$$\begin{cases} \dot{I}^{el}_{D(DM)} = -0.5\left(\dot{I}^{el}_{D(DJ)} + \dot{I}^{el}_{D(DK)}\right) \\ \dot{I}^{fl}_{D(DM)} = -0.5\left(\dot{I}^{fl}_{D(DJ)} + \dot{I}^{fl}_{D(DK)}\right) \\ \dot{I}^{gl}_{D(DM)} = -0.5\left(\dot{I}^{el}_{D(DJ)} - \dot{I}^{el}_{D(DK)}\right) \\ \dot{I}^{hl}_{D(DM)} = -0.5\left(\dot{I}^{fl}_{D(DJ)} - \dot{I}^{fl}_{D(DK)}\right) \end{cases}$$
(7)

The interface equations of point D, Line DJ are constructed by Lines DM and DK:

$$\begin{cases} \dot{U}^{el}_{D(DJ)} = 2\dot{U}^{el}_{D(DM)} - \dot{U}^{el}_{D(DK)} \\ \dot{U}^{fl}_{D(DJ)} = 2\dot{U}^{fl}_{D(DM)} - \dot{U}^{fl}_{D(DK)} \end{cases}$$
(8)

$$\begin{cases} \dot{I}^{el}_{D(DJ)} = -2\dot{I}^{el}_{D(DM)} - \dot{I}^{el}_{D(DK)} \\ \dot{I}^{fl}_{D(DJ)} = -2\dot{I}^{fl}_{D(DM)} - \dot{I}^{fl}_{D(DK)} \end{cases}$$
(9)

The interface equations of point D, Line DK are constructed by Lines DM and DJ:

$$\begin{cases} \dot{U}^{el}_{D(DK)} = 2\dot{U}^{el}_{D(DM)} - \dot{U}^{el}_{D(DJ)} \\ \dot{U}^{fl}_{D(DK)} = 2\dot{U}^{fl}_{D(DM)} - \dot{U}^{fl}_{D(DJ)} \end{cases}$$
(10)

$$\begin{cases} \dot{I}^{el}_{D(DK)} = -2\dot{I}^{el}_{D(DM)} - \dot{I}^{el}_{D(DJ)} \\ \dot{I}^{fl}_{D(DK)} = -2\dot{I}^{fl}_{D(DM)} - \dot{I}^{fl}_{D(DJ)} \end{cases}$$
(11)

4 FAULT REGION DETERMINATION METHOD OF PARTIAL COUPLING FOUR-CIRCUIT TRANSMISSION LINES

4.1 Determination criteria of fault region

If a fault occurs in Line DM, $\dot{U}_{D(J)}^{ml} = \dot{U}_{D(DJ)}^{ml}$ and $\dot{U}_{D(K)}^{ml} = \dot{U}_{D(DK)}^{ml}$. Combined with Eq. (6), we can obtain:

$$\begin{cases} \left|\dot{U}_{D(M)}^{el}\right| < \left|\dot{U}_{D(DM)}^{el}\right| = 0.5\left|\dot{U}_{D(J)}^{el} + \dot{U}_{D(K)}^{el}\right| \\ \left|\dot{U}_{D(M)}^{f1}\right| > \left|\dot{U}_{D(DM)}^{f1}\right| = 0.5\left|\dot{U}_{D(J)}^{f1} + \dot{U}_{D(K)}^{f1}\right| \\ \left|\dot{U}_{D(M)}^{g1}\right| > \left|\dot{U}_{D(DM)}^{g1}\right| = 0.5\left|\dot{U}_{D(J)}^{el} - \dot{U}_{D(K)}^{el}\right| \\ \left|\dot{U}_{D(M)}^{h1}\right| > \left|\dot{U}_{D(DM)}^{h1}\right| = 0.5\left|\dot{U}_{D(J)}^{f1} - \dot{U}_{D(K)}^{f1}\right| \end{cases} \quad (12)$$

If a fault occurs in Line DJ, we can obtain:

$$\begin{cases} \left|\dot{U}_{D(M)}^{ml}\right| = \left|\dot{U}_{D(DM)}^{ml}\right| \quad (m=e,f,g,h) \\ \left|\dot{U}_{D(K)}^{ml}\right| = \left|\dot{U}_{D(DK)}^{ml}\right| \quad (m=e,f) \\ \left|\dot{U}_{D(J)}^{el}\right| < \left|\dot{U}_{D(DJ)}^{el}\right| \\ \left|\dot{U}_{D(J)}^{f1}\right| > \left|\dot{U}_{D(DJ)}^{f1}\right| \end{cases} \quad (13)$$

From Eqs. (6) and (15), we can obtain:

$$\begin{cases} \left|\dot{U}_{D(M)}^{el}\right| = \left|\dot{U}_{D(DM)}^{el}\right| > 0.5\left|\dot{U}_{D(J)}^{el} + \dot{U}_{D(K)}^{el}\right| \\ \left|\dot{U}_{D(M)}^{f1}\right| = \left|\dot{U}_{D(DM)}^{f1}\right| < 0.5\left|\dot{U}_{D(J)}^{f1} + \dot{U}_{D(K)}^{f1}\right| \\ \left|\dot{U}_{D(M)}^{g1}\right| = \left|\dot{U}_{D(DM)}^{g1}\right| > 0.5\left|\dot{U}_{D(J)}^{el} - \dot{U}_{D(K)}^{el}\right| \\ \left|\dot{U}_{D(M)}^{h1}\right| = \left|\dot{U}_{D(DM)}^{h1}\right| < 0.5\left|\dot{U}_{D(J)}^{f1} - \dot{U}_{D(K)}^{f1}\right| \end{cases} \quad (14)$$

If a fault occurs in Line DK, similarly, Eq. (14) can be obtained.

By comparison, from Eqs. (12) and (14), no matter whether a fault occurs in Line DM, the symbols of g modulus inequality are the same. Consequently, g modulus inequality cannot be used in fault region determination. But the symbols of e, f and h modulus inequality relate to fault regions, so that they can be used in fault region determination. The criterion that a fault occurs in Line DM is as follows:

$$\begin{cases} \Delta U_e = 0.5\left|\dot{U}_{D(J)}^{el} + \dot{U}_{D(K)}^{el}\right| - \left|\dot{U}_{D(M)}^{el}\right| > 0 \\ \Delta U_f = \left|\dot{U}_{D(M)}^{f1}\right| - 0.5\left|\dot{U}_{D(J)}^{f1} + \dot{U}_{D(K)}^{f1}\right| > 0 \\ \Delta U_h = \left|\dot{U}_{D(M)}^{h1}\right| - 0.5\left|\dot{U}_{D(J)}^{f1} - \dot{U}_{D(K)}^{f1}\right| > 0 \end{cases} \quad (15)$$

If the fault region is not Line DM, from Eq. (5), the true values $\dot{U}_{D(DM)}^{el}$ and $\dot{U}_{D(DM)}^{g1}$ of modulus e and g can be deduced from terminal M.

From Eq. (6), we can obtain:

$$\begin{cases} \dot{U}_{D(DJ)}^{el} = \dot{U}_{D(DM)}^{el} + \dot{U}_{D(DM)}^{g1} \\ \dot{U}_{D(DK)}^{el} = \dot{U}_{D(DM)}^{el} - \dot{U}_{D(DM)}^{g1} \end{cases} \quad (16)$$

Because equal directional voltage of fault line is smaller than that of the perfect line, the assistant criterion that a fault occurs in Line DJ is:

$$\left|\dot{U}_{D(DJ)}^{el}\right| < \left|\dot{U}_{D(DK)}^{el}\right| \quad (17)$$

If it cannot accord with Eqs. (15) and (17), the fault occurs in Line DK.

4.2 Criterion screening conditions

When faults across lines occur in the phases with the same name, at least one of the three circulating components does not exist. Then, one of the circulating component criteria in Eq. (15) is out of work. Especially, when fault occurs in four lines at the same time, three criteria are all out of work. At this moment, only equal directional component criteria are effective. Therefore, to avoid the interference of failure criteria, three above criteria must be screened in advance in order to choose the most effective criterion.

When faults across lines occur in the phases with the same name, at least one of the three circulating components does not exist, namely the circulating component with minimal voltage. The differences before and after the fault are small. Therefore, the minimal circulating component criteria may cause misjudgment easily, so that it must be weeded out. The circulating voltage of point D before the fault can be regarded as the reference value to measure circulating component. Here, the circulating voltage of point D before the fault can be achieved by circulating current of terminal M based on Eq. (3), and that after the fault is the larger one between $\left|\dot{U}_{D(M)}^{f1}\right|$ and $0.5\left|\dot{U}_{D(J)}^{f1} + \dot{U}_{D(K)}^{f1}\right|$ in Eq. (15). Respectively set $U_{set.f}$ and $U_{set.h}$ as 1.5 times of f and h modulus voltage in point D and Line DM before the fault, and we can obtain:

$$\max\left(\left|\dot{U}_{D(M)}^{f1}\right|, 0.5\left|\dot{U}_{D(J)}^{f1} + \dot{U}_{D(K)}^{f1}\right|\right) < U_{set.f} \quad (18)$$

$$\max\left(\left|\dot{U}_{D(M)}^{h1}\right|, 0.5\left|\dot{U}_{D(J)}^{f1} - \dot{U}_{D(K)}^{f1}\right|\right) < U_{set.h} \quad (19)$$

If it accords with Eq. (18), circulating component f is very small, thus weeding out circulating component f criterion in Eq. (15); if it accords with Eq. (19), circulating component h criterion in Eq. (15) can be weeded out.

In addition, if a fault occurs near point D, all criteria values are close to 0, and the sensitivity is very low. In view of this, the sensitivity of each criterion should be assessed in order to choose the modulus criterion with the highest sensitivity for judgment. According to Eq. (15), the larger the difference between $0.5|\dot{U}_{D(J)}^{m1}+\dot{U}_{D(K)}^{m1}|$ and $|\dot{U}_{D(M)}^{m1}|$ is, the more sensitive the criterion is. Define $K_{Sen.m}$ as the difference degree between two values of modulus m criterion in Eq. (15):

$$K_{Sen.m}=\frac{\max\left(\left|\dot{U}_{D(M)}^{m1}\right|,\,0.5\left|\dot{U}_{D(J)}^{m1}+\dot{U}_{D(K)}^{m1}\right|\right)}{\min\left(\left|\dot{U}_{D(M)}^{m1}\right|,\,0.5\left|\dot{U}_{D(J)}^{m1}+\dot{U}_{D(K)}^{m1}\right|\right)} \quad (20)$$

where $m = e, f, h$. The larger $K_{Sen.m}$ is, the more sensitive the criterion is. Thus, we can choose the modulus criterion with the largest value of $K_{Sen.m}$ for judgment.

5 FAULT LOCATION OF THREE-TERMINAL PARTIAL COUPLING FOUR-CIRCUIT TRANSMISSION LINES

When the fault region is confirmed, based on the long line equation, fault location equations of three terminal partial coupling four-circuit transmission lines can be achieved.

If the fault region is Line DM, namely the fault occurs at a distance of x to terminal M, modulus voltage and current of point D and Line DM can be obtained from Eqs. (4), (5), (6) and (7). Based on the voltage and current of each terminal to the fault point, and considering the same voltage of both sides of the fault point, the fault location equation can be obtained:

$$\dot{U}_M^{m1}ch(\gamma_{DM}^1x)-\dot{I}_M^{m1}Z_{DM}^1sh(\gamma_{DM}^1x)$$
$$=\dot{U}_{D(DM)}^{m1}ch[\gamma_{DM}^1(L_{DM}-x)]$$
$$-\dot{I}_{D(DM)}^{m1}Z_{DM}^1sh[\gamma_{DM}^1(L_{DM}-x)] \quad (m=e,f,g,h)$$
$$(21)$$

If the fault region is Line DJ, namely the fault occurs at a distance of x to terminal J, modulus voltage and current of point D and Line DJ can be obtained from Eqs. (3), (5), (8) and (9). The fault location equation is:

$$\dot{U}_J^{m1}ch(\gamma_{DJ}^1x)-\dot{I}_J^{m1}Z_{DJ}^1sh(\gamma_{DJ}^1x)$$
$$=\dot{U}_{D(DJ)}^{m1}ch[\gamma_{DJ}^1(L_{DJ}-x)]$$
$$-\dot{I}_{D(DJ)}^{m1}Z_{DJ}^1sh[\gamma_{DJ}^1(L_{DJ}-x)] \quad (m=e,f) \quad (22)$$

If the fault region is Line DK, namely the fault occurs at a distance of x to terminal K, modulus voltage and current of point D and Line DK can be obtained from Eqs. (3), (4), (10) and (11). The fault location equation is:

$$\dot{U}_K^{m1}ch(\gamma_{DK}^1x)-\dot{I}_K^{m1}Z_{DK}^1sh(\gamma_{DK}^1x)$$
$$=\dot{U}_{D(DK)}^{m1}ch[\gamma_{DK}^1(L_{DK}-x)]$$
$$-\dot{I}_{D(DK)}^{m1}Z_{DK}^1sh[\gamma_{DK}^1(L_{DK}-x)] \quad (m=e,f) \quad (23)$$

After the fault region is determined, the fault distance can be obtained through solving the corresponding fault location equations. Thus, fault location algorithm of three-terminal partial coupling four-circuit transmission lines can be achieved.

6 SIMULATION, ANALYSIS AND VERIFICATION

To verify the fault location algorithm of three-terminal partial coupling four-circuit transmission lines, this paper uses ATP/EMTP to construct a model of three-terminal partial coupling four-circuit transmission lines as Figure 1 shows. The line model uses J. Marti frequency dependent model. The voltage level is 220 KV. The lengths of each line are as follows: L_{DM} = 100 km, L_{DJ} = 100 km and L_{DK} = 80 km. The fault simulations for different types of fault and transition resistance are set in different fault positions of Line DM and DJ. Firstly, the phasor filter algorithm proposed in Ref. [8] is employed to extract power frequency phasor of each phase voltage and current, which can eliminate the influence of decaying DC component and increase the accuracy of power frequency fault component. Then, the fault location algorithm proposed in this paper and the traditional double-terminal fault location method that ignores the influence of partial line coupling are respectively used to calculate the fault location. Table 1 and Table 2 show the fault location consequence of metallic fault and high resistance single-phase ground fault (200 Ω), respectively. In Table 1 and 2, the observation point of fault position and fault location consequence is terminal M, and coupling demarcation point D is located at a distance of 100 km to terminal M. IAG denotes A-phase ground fault in line-I, IAIIBCG denotes A-phase ground fault in line-I and BC-phase ground fault in line-II, and so on.

From Table 1, for the metallic ground fault, the maximum error of the traditional method is 1.31 km, while that of our proposed method is 0.69 km. Besides, the fault region determination consequences are all correct in the proposed method.

Table 1. Results of fault location of metallic fault.

Fault position		Fault type	Region determination result	Proposed method		Traditional method	
				X (km)	Error (km)	X (km)	Error (km)
The distance to terminal M (Line DM)	1 km	IAG	DM	0.74	−0.26	0.66	−0.34
		IABG	DM	0.75	−0.25	0.68	−0.32
		IABC	DM	0.77	−0.23	0.7	−0.3
		IAIIAG	DM	0.74	−0.26	0.65	−0.35
		IAIIBCG	DM	0.77	−0.23	0.7	−0.3
	10 km	IAG	DM	9.49	−0.51	9.5	−0.5
		IABG	DM	9.55	−0.45	9.57	−0.43
		IABC	DM	9.58	−0.42	9.59	−0.41
		IAIIAG	DM	9.5	−0.5	9.5	−0.5
		IAIIBCG	DM	9.59	−0.41	9.59	−0.41
	50 km	IAG	DM	49.49	−0.51	49.7	−0.3
		IABG	DM	49.31	−0.69	49.44	−0.56
		IABC	DM	49.36	−0.64	49.42	−0.58
		IAIIAG	DM	49.47	−0.53	49.69	−0.31
		IAIIBCG	DM	49.59	−0.41	49.73	−0.27
	90 km	IAG	DM	89.74	−0.26	89.6	−0.4
		IABG	DM	89.91	−0.09	89.7	−0.3
		IABC	DM	89.86	−0.14	89.63	−0.37
		IAIIAG	DM	89.72	−0.28	89.55	−0.45
		IAIIBCG	DM	89.9	−0.1	89.65	−0.35
	99 km	IAG	DM	98.92	−0.08	98.34	−0.66
		IABG	DM	98.97	−0.03	97.74	−1.26
		IABC	DM	98.98	−0.02	97.69	−1.31
		IAIIAG	DM	98.88	−0.12	98.22	−0.78
		IAIIBCG	DM	99.02	0.02	98.24	−0.76
The distance to terminal M (Line DJ)	101 km	IAG	DJ	100.84	−0.16	100.71	−0.29
		IABG	DJ	101.04	0.04	101.46	0.46
		IABC	DJ	100.83	−0.17	101.23	0.23
		IAIIAG	DJ	100.35	−0.65	100.62	−0.38
		IAIIBCG	DJ	100.84	−0.16	100.71	−0.29
	110 km	IAG	DJ	109.55	−0.45	109.63	−0.37
		IABG	DJ	110.01	0.01	109.99	−0.01
		IABC	DJ	109.92	−0.08	109.87	−0.13
		IAIIAG	DJ	109.53	−0.47	109.44	−0.56
		IAIIBCG	DJ	109.91	−0.09	109.39	−0.61
	150 km	IAG	DJ	150.06	0.06	150.44	0.44
		IABG	DJ	150.53	0.53	150.56	0.56
		IABC	DJ	150.4	0.4	150.41	0.41
		IAIIAG	DJ	150.08	0.08	150.31	0.31
		IAIIBCG	DJ	150.39	0.39	150.16	0.16
	190 km	IAG	DJ	190.31	0.31	190.5	0.5
		IABG	DJ	190.58	0.58	190.59	0.59
		IABC	DJ	190.46	0.46	190.45	0.45
		IAIIAG	DJ	190.31	0.31	190.45	0.45
		IAIIBCG	DJ	190.45	0.45	190.37	0.37
	199 km	IAG	DJ	198.99	−0.01	199.05	0.05
		IABG	DJ	199.07	0.07	199.05	0.05
		IABC	DJ	199.07	0.07	199.04	0.04
		IAIIAG	DJ	198.99	−0.01	199.05	0.05
		IAIIBCG	DJ	199.05	0.05	198.99	−0.01

Table 2. Results of fault location of high resistance.

Fault position		Fault type	Region determination result	Proposed method		Traditional method	
				X (km)	Error (km)	X (km)	Error (km)
The distance to terminal M (Line DM)	1 km	IAG	DM	1.01	0.01	2	1
		IABG	DM	0.74	−0.26	0.92	−0.08
		IABC	DM	0.78	−0.22	0.45	−0.55
		IAIIAG	DM	0.9	−0.1	0.74	−0.26
		IAIIBCG	DM	0.77	−0.23	0.46	−0.54
	10 km	IAG	DM	9.82	−0.18	7.68	−2.32
		IABG	DM	9.15	−0.85	8.48	−1.52
		IABC	DM	9.18	−0.82	8.69	−1.31
		IAIIAG	DM	9.8	−0.2	8.7	−1.3
		IAIIBCG	DM	9.74	−0.26	9.01	−0.99
	50 km	IAG	DM	49.67	−0.33	48.42	−1.58
		IABG	DM	49.34	−0.66	49.07	−0.93
		IABC	DM	49.32	−0.68	49.16	−0.84
		IAIIAG	DM	49.47	−0.53	49.69	−0.31
		IAIIBCG	DM	49.59	−0.41	49.73	−0.27
	90 km	IAG	DM	89.71	−0.29	90.24	0.24
		IABG	DM	89.9	−0.1	90.07	0.07
		IABC	DM	89.86	−0.14	89.63	−0.37
		IAIIAG	DM	89.66	−0.34	89.83	−0.17
		IAIIBCG	DM	89.79	−0.21	89.83	−0.17
	99 km	IAG	DM	98.79	−0.21	100.44	1.44
		IABG	DM	98.85	−0.15	99.35	0.35
		IABC	DM	98.76	−0.24	98.83	−0.17
		IAIIAG	DJ	98.35	−0.65	99.23	0.23
		IAIIBCG	DM	98.83	0.17	98.89	−0.11
The distance to terminal M (Line DJ)	101 km	IAG	DJ	100.14	−0.86	101.3	0.3
		IABG	DJ	101.16	0.16	101.06	0.06
		IABC	DJ	100.98	−0.02	100.97	−0.03
		IAIIAG	DJ	100.29	−0.71	100.94	−0.06
		IAIIBCG	DJ	100.87	−0.13	100.91	−0.09
	110 km	IAG	DJ	109.48	−0.52	111.68	1.68
		IABG	DJ	110.23	0.23	110.06	0.06
		IABC	DJ	110.06	0.06	110	0
		IAIIAG	DJ	109.47	−0.53	110.13	0.13
		IAIIBCG	DJ	110.05	0.05	110.1	0.1
	150 km	IAG	DJ	150.2	0.2	152.28	2.28
		IABG	DJ	150.43	0.43	150.61	0.61
		IABC	DJ	150.37	0.37	150.43	0.43
		IAIIAG	DJ	150.21	0.21	151.32	1.32
		IAIIBCG	DJ	150.49	0.49	150.79	0.79
	190 km	IAG	DJ	190.22	0.22	192.36	2.36
		IABG	DJ	190.5	0.5	190.95	0.95
		IABC	DJ	190.46	0.46	190.73	0.73
		IAIIAG	DJ	190.34	0.34	191.51	1.51
		IAIIBCG	DJ	190.46	0.46	191.05	1.05
	199 km	IAG	DJ	199.02	0.02	201.78	2.78
		IABG	DJ	199.05	0.05	199.87	0.87
		IABC	DJ	199.07	0.07	199.6	0.6
		IAIIAG	DJ	199.01	0.01	200.45	1.45
		IAIIBCG	DJ	199.05	0.05	199.91	0.91

From Table 2, for the high resistance ground fault, the error is higher than 1 km many times and the maximum error is 2.78 km in the traditional method, which dissatisfies the accuracy requirement of fault location. However, in the proposed algorithm, the maximum error is only 0.86 km. In addition, when IAIIAG occurs at a distance of 99 km to terminal M, although fault region is misjudged in Line DJ, the error is only 0.65 km. At a distance of 99 km to 110 km to terminal M, the fault location error is not more than 0.86 km no matter what the fault region determination consequence is.

7 CONCLUSIONS

In summary, the interface model of coupling demarcation point constructed in this paper reflects the fault analysis network characteristics of demarcation point, and the proposed criteria are effective to determine the fault region. Based on the proposed fault location algorithm, a great transition resistance is allowed, and a high accuracy of fault location can be achieved.

REFERENCES

[1] Li Qiang, Wang Yinle, Fault Location Algorithm of HV Transmission Line [J]. Power System Protection and Control. 2009, 37(23):192–197.
[2] Qin Jian, Peng Liping, Wang Hechun. Single terminal methods of traveling wave fault location in transmission line using wavelet transform [J]. Automation of Electric Power Systems. 2005, 29(19):62–65.
[3] Suonan Jiale, Zhang Yining, Qi Jun, Jiao Zaibin. Time Domain Fault Location Method Based on Transmission Line Parameter Identification Using Two Terminals Data [J]. Power System Technology, 2006, 30(8):65–70.
[4] Zhang Xiaoming, Xu Yan, Wang Yu, et al. A fault location algorithm for two-terminal transmission lines based on parameter detection [J]. Power System Protection and Control, 2011, 39(12):106–111.
[5] Luo Xiantong, Luo Jian. A time-domain fault location algorithm for high resistance grounding using one terminal data [J]. Power System Protection and Control, 2010, 38(8):6–9.

[6] Liang Yuansheng, Wang Gang, LI Haifeng. Fault Location Algorithm Based on Time-frequency-domain with Two Terminals Asynchronou [J]. Automation of Electric Power Systems, 2009, 33(4):62–66.
[7] Zhao Yongxian, Cao Xiaoguai, Liu Wangshun. New fault location algorithm for double transmission lines on a same tower [J]. Automation of Electric Power Systems, 2005, 29(4):72–76.
[8] Gong Zhendong, Fan Chunju, Yu Weiyong, A new fault allocation algorithm for parallel transmission line based on six sequence network. Automation of Electric Power Systems, 2007, 31(17):58–63.
[9] Song Guobing, Suonan Jiale, Xu Qingqiang, et al. Parallel transmission lines fault location algorithm based on differential component net. IEEE Trans. on Power Delivery, 2005, 20(4):2396–2406.
[10] Xu Peng, Liang Yuansheng, Wang Gang. A parameter adaptive fault location with two terminal data for four parallel transmission lines on the same tower. Automation of Electric Power Systems. 2010, 34(9):59–64.
[11] Gong Zhendong, Fan Chunju, Tian Yu. A fault location algorithm suitable for jointed four transmission lines. Automation of Electric Power Systems, 2007, 31(23):70–73.
[12] Ma Jing, Shi Yuxin, Ma Wei, Wang Zengping. Distributed Parameter Based One-End Fault Location for Inter-Line Fault and Earth Fault in Double-Circuit Transmission Lines on Same Tower [J]. Power System Technology. 2014, (9):2525–2531.
[13] Tian Yu. Study on fault location of parallel transmission teed line [D]. Shanghai Jiao Tong University, 2007.
[14] Tian Yu, Fan Chunju, Gong Zhendong. Faulted line selecting method of T cireuit of parallel transmission Lines [J]. Automation of Electric Power Systems, 2006, 30(21):71–76.
[15] Tian Yu, Fan Chunju, Gong Zhendong, Li Shuai. A Faulted Line Selection Method of Parallel Transmission Teed Line on the Basis of Differential Current [J]. Automation of Electric Power Systems, 2007, 31(3):67–71.
[16] Liang Yuansheng, Wang Gang, Li Haifeng. A filtering algorithm for eliminating effect of transient component and its application on fault location. Automation of Electric Power Systems. 2007, 31(22):77–82.

Electrical, Control Engineering and Computer Science – Liu (Ed.)
© *2016 Taylor & Francis Group, London, ISBN 978-1-138-02937-8*

Online identification based stability index in Multi-Send HVDC power transmission system

Shuang Zhang, Feng Gao & Bei Tian
Power Science Research Institute of State Grid Ningxia Power Co., Yinchuan, China

Yinfeng Wang & Chao Lu
Tsinghua University, Beijing, China

ABSTRACT: With the development of HVDC, some typical Multi-Send HVDC systems (MSDC) appears, and new voltage and frequency problems are becoming the threats to system stability. In this paper, Multi-infeed Effective Short Circuit Ratio (MESCR) is selected as a general index to evaluate the stable degree in Multi-send HVDC system. Ambient signals in power system containing lots of important information are used to identify the sensitivity among different node voltages, and MESCR is calculated based on the identified sensitivities. An online identification method is also proposed to meet the demand of real-time online monitoring. A benchmark MSDC simulation system is built to validate the proposed identification method and stability index.

Keywords: Multi-Send HVDC system; stability index; ambient signals; sensitivity analysis; online identification method

1 INTRODUCTION

Construction of large-scale interconnected grid is helpful to improve the overall stability of power systems, and to solve the problems that power centers and load centers are not coincident in China. At present, lots of large energy bases are continuously constructed in the southwest, northwest and some other places of China, where power resources are abundant. So large quantities of power needs to be transmitted to areas with industrial aggregation and large population. Several HVDC transmissions are built radially outwards from the energy centers. Thus, systems with HVDC lines from the same AC system are called Multi-Send HVDC systems (MSDC).

It is widely recognized that Multi-infeed Effective Short Circuit Ratio (MESCR) index is adopted to evaluate the power transmitting ability and the voltage supporting strength in AC-DC power systems. Though there are several forms of this index, neither of them can realize the objective of online identification. However, a certain form of MESCR index based on the Multi-Infeed Interaction Factor (MIIF) proposed by CIGRE is chosen in this paper. Then a node voltage sensitivity identification model is established, and the ambient signals is real-timely collected from the system to calculate the MIIF or rather a sensitivity factor.

2 MESCR STABILITY INDEX BASED ON MIIF

It is generally accepted that the ratio between short circuit capacity of converter bus and DC power is a manifestation of stability in AC-DC systems. This ratio is defined as Short Circuit Ratio (SCR) index. It is widely used to evaluate the stability of a single DC system, and has been proved to be effective in practical applications. Analogously, this ratio can be also applied to Multi-DC systems, but mutual influence between the DC transmissions and compensation of reactive power at converter bus should be taken into account. Therefore, MESCR, an index that includes the above factors based on MIIF is defined as Eq. 1:

$$MESCR_i = \frac{S_{aci} - Q_{ci}}{P_{di} + \sum_{j=1, j \neq i}^{n} (MIIF_{ji} \times P_{dj})} \tag{1}$$

where i is the number of converter bus under consideration; S_{aci} is the short circuit capacity of the ith DC transmission; Q_{ci} is the reactive shunt compensation at the ith converter bus; P_{di}, P_{dj} are the power of the ith and the jth converter station respectively.

In Eq. 1, the participation index is MIIF that takes the influence of other converters connected to the concerned bus into account. MIIF is known as an empirical and experimental indicator, which is proposed as the ratio between the voltage changes at converter bus j and bus i due to a reactive power change at the bus i. In a practical application, to ensure that the voltage changes are small and do not change the system operation mode, it is suggested that the quantum of reactive power change caused by switching electric reactors at bus i should be only about 1% amplitude of the nominal voltage.

According to Eq. 1, the parameters, such as S_{aci}, Q_{ci}, P_{di} and P_{dj}, can be measured dynamically and real-timely in a real power system. But the traditional MIIF index is restricted to simulations or site trials, and hard to be calculated continuously in the real-time mode. Now, reference to the sensitivity analysis method based on the ambient signals, a new similar index is put forward in the following section, which will make it available that the MESCR index can be evaluated rapidly and real-timely.

3 MESCR CALCULATION BASED ON ONLINE IDENTIFICATION

Power system is a complicated process that has strong nonlinearity, which can be expressed by a series of differential-algebraic equations as Eq. 2.

$$\begin{cases} \dot{x} = f(x, y, u, p) \\ 0 = g(x, y, u, p) \end{cases} \tag{2}$$

where x is state vector, including electric generators' power-angle, rotate speed and so on. y is algebraic vector, including buses voltage, nodes' current and so on; u is input vector, including command of the controllers, output of the prime motors; p includes some parameters about operation modes, such as transmission line parameters and generators' power, etc.

When the power system operates in static stability or just a small perturbance occurs in the power system, the system state equations can be simplified at the equilibrium points of power system based on some linearized assumptions as Eq. 3.

$$\begin{cases} \Delta \dot{x} = A(p)\Delta x + B(p)\Delta u \\ \Delta y = C(p)\Delta x + D(p)\Delta u \end{cases} \tag{3}$$

where A is the state matrix; B is the input matrix; C is the output matrix; D is the transfer matrix. And the elements in A, B, C and D are related to p— the parameters of operation mode.

Under the condition of steady state, the relation between the voltage of converter bus and the other buses around can be used to evaluated their interactions, referring to the definition of MIIF. According to the Eq. 3, the relation between the converter bus voltage can be linearly expressed when the system is statically stable or small-signally stable. If the voltage of the converter bus is choose as a state variable, and the others are chosen as several input variables. The mathematical relationship can be expressed as Eq. 4 at every short period of sampling and calculating.

$$\Delta U_i = \sum_{j=1, j \neq i}^{k} \frac{\partial U_i}{\partial U_j} \Delta U_j + \varepsilon_i. \tag{4}$$

where ΔU_i is the voltage change of converter bus; ΔU_j is the voltage changes of the buses around the converter bus; ε_i is the residual caused by approximate description. $\partial U_i / \partial U_j$ is the sensitivity of ΔU_i relative to ΔU_j, which is similar to MIIF, but includes more factors that affects the interactions. In order to simplify the discussion, this indicator is defined as Nodal Voltage Sensitivity Index (NVSI).

Obviously, it is a multiple linear regression model. In this model, ΔU_i ($i = 1, 2, 3...$) are random variables. They are not the voltage changes made by reactors, but voltage ambient signals that are inherent fluctuation signals in power system. Ambient signals are generally considered as a kind of white noise signal. Thus the model becomes a Gauss–Markov linear regression model.

The ambient signals caused by switching equipment, load changes and other factors are real-timely collected through PMU in actual electrical networks, so the NVSI can be calculated rapidly and continuously after a suitable process. Furthermore, the new stability index of MSDC systems in the Eq.5 can also be identified online.

$$MESCR_i^* = \frac{S_{aci} - Q_{ci}}{P_{di} + \sum_{j=1, j \neq i}^{n} (NVSI_{ji} \times P_{dj})}. \tag{5}$$

The key issue is that the $MESCR^*$ must be calculated in a short time (Δt), and the system's operation mode should be considered to keep changeless. While the more sampling points are, the more accurate regression results is. So there is a balance between speed and accuracy.

4 SIMULATION RESULTS

To verify the reliability of the index and the online identification method above, A two-send HVDC simulation system based on the standard CEPRI

7 nodes system is constructed . The structure diagram of this simulation model is shown in the Figure 1.

The generators, DC transmissions, loads and so on in the simulation system are all set with parameters as the original model. Then a check is made to make sure the reliability and stability of the system. Flow of the system should be converged and reasonable. The system further should operate with sufficient transient stability margin.

In order to simulate the ambient signals in actual systems, some Gaussian white noise signals are injected into some appropriate nodes. The contrast of ambient signals' power spectrum between the simulation system and an actual system is presented in Figure 2 and Figure 3.

According to Figures 2 and 3, the ambient signals in the simulation system and actual system both distribute in a similar frequency range. Then the simulation system's stability degree is changed by adjusting the network structure, such as the transmission line impedances, the size of loads, etc. Because the sampling rate of PUM is about 100/s in power system, the simulation step is 0.01 s.

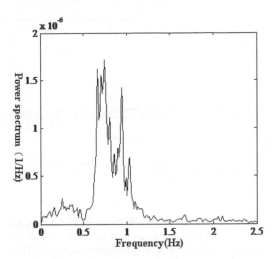

Figure 3. Power spectrum of ambient signals in actual system.

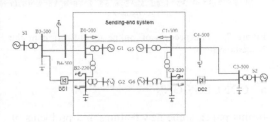

Figure 1. A two-send HVDC simulation system.

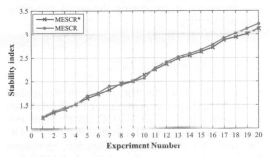

Figure 4. MESCR and MESCR* for different levels of stability.

When the system operation mode doesn't change, there is a contrast of stability index got separately by methods based on MIIF and NVSI in Figure 4. Obviously, there is little difference between the two methods, so the stability evaluated by sensitivity analysis is effective in a multi-send HVDC system.

Then, a simulation of the online identification in the simple system is made. In the sampling process, an overlapping method with rectangle window is used to ensure the continuity of identification results. Moreover, the linear stepwise regressive method is applied to remove the irrelevant buses in Figure 5.

The system's stability was changed at t = 4 min, simply caused by a three-phase break fault at the AC line of C1–500 to C4–500. The stability indexes of both DC systems decreased, and the stability of DC2 went down even more, because it was closer to the fault point than DC1.

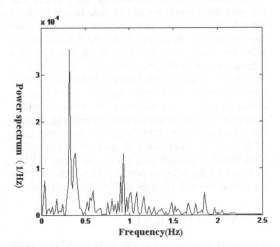

Figure 2. Power spectrum of ambient signals in simulation system.

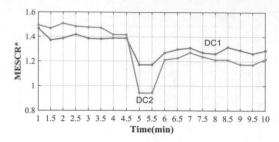

Figure 5. The process of online identification in the two-send HVDC system.

According the Figure 5, the new stability index and online identification can properly reflect the static stability of the simulation system. They can also correctly show some slower changes of stability in the systems. This online identification method has great potential in stability monitoring and early risk warning. Certainly, further studies concentrating on improving the speed and accuracy are worthy.

5 ACTUAL SYSTEM TEST

Further, the online identification method is applied in the Ningxia power grid planned in 2016, which is a typical multi-send HVDC system with two HVDC transmissions. One is the ±660 KV HVDC line from Yinchuandong to Qingdao, Shandong Province, the other is the ±800 KV HVDC line from Taiyangshan to Shaoxing, Zhejiang Province, which is completed in the future. The geographical position of the two converter station is shown in Figure 6. The electrical distance of two stations is close, so that the interaction between them may cause some stability problems. The online identification method can be used to real-timely monitor the stability indexes of the two HVDC systems.

The same solution as in the above simulation is taken that the system stability is changed at t = 4 min by a three-phase break fault at three AC lines around the Yinchuandong converter station (330 KV Yinchuandong to Tianshuihe). The phenomena that appear in the the system can be reflected by the stability index based on the NVSI in Figure 7.

Similarly, the process of stability index identified online can better reflects the changes of stability caused by operation mode variations. According to the different values of change, a preliminary judgment of the failures' position and could be made.

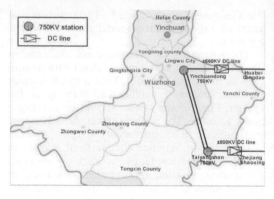

Figure 6. Geographical position of HVDC lines in Ningxia Province.

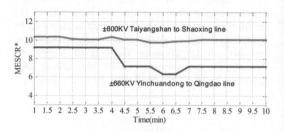

Figure 7. The process of online identification in the Ningxia system.

6 CONCLUSIONS

In this paper, some new features and problems of multi-send HVDC system are discussed. Then a multiple linear regression model based on the sensitivity analysis method is built on the premise that the operation mode doesn't change in a very short time. Furthermore, a new stability index that consists of nodal voltage sensitivity index is defined to solve the defect that the traditional index need a tedious process and cannot be identified real-timely. An advantage of the new index is that the identification only needs the sampled data of ambient signals rather than the parameters of the power network structure, so it is an attempt to the make full use of the data sampled by PMUs all over the power grid. At the last two sections, a simulation of the online identification is made in a simple two-send HVDC system and the Ningxia power grid, which show the practicability and reliability of the new method.

ACKNOWLEDGMENT

This work was supported in part by Major State Basic Research Development Program of China

(2012CB215206) and National Natural Science Foundation of China (51037002 and 51107061).

REFERENCES

CIGRE Working Group B4.41. 2008. *"Systems with multiple DC infeed"*, Paris: CIGRE.

Denis L.H.A & Andersson G. 1998. "Power stability analysis of multi-infeed HVDC systems". *IEEE Trans. Power Delivery*, Vol. 13, pp. 923–931.

Denis L.H.A & Andersons G. 1998. "Use of Participation factors in modal voltage stability analysis of multi-infeed HVDC systems", *IEEE Trans. Power Delivery*, Vol. 13, pp. 203–211.

Gavrilovic. 1991. "AC/DC system strength as indicated by short circuit ratios. in AC and DC Power Transmission", International Conference on. London.

Kundur P. 1994. *"Power System Stability and Control"*. New York: McGraw-Hill.

Nayak, R.N, et al, 2006. "AC/DC interactions in multi-infeed HVDC scheme: a case study", in Power India Conference, IEEE.

Paulo Fischer & De Toledo, B.B.G.A. 2005. "Multiple infeed short circuit ratio—Aspects multiple HVDC into one AC network", Transmission and Distribution Conference and Exhibition: Asia and Pacific, IEEE/PES, pp. 1–6.

Weifang Lin & Tang Yong, et al. 2010. "Voltage Stability Analysis of Multi-in feed AC/DC Power System Based on Multi-infeed Short Circuit Ratio", Power System Technology, 2010 International Conference on, Hangzhou.

Yao W., & L. Jiang, et al. 2011. "Delay-Dependent Stability Analysis of the Power System With a Wide-Area Damping Controller Embedded", *IEEE Transactions on power systems*, Vol. 26, No. 1, pp. 230–240.

Yong Li. & Christian Rehtanz, et al. 2012. "Assessment and Choice of Input Signals for Multiple HVDC and FACTS Wide-Area Damping Controllers", *IEEE Transactions on power systems*, Vol. 27, No. 4, pp. 1969–1977.

Electrical, Control Engineering and Computer Science – Liu (Ed.)
© *2016 Taylor & Francis Group, London, ISBN 978-1-138-02937-8*

Stability analysis of hybrid AC/DC micro-grid

Hao Pan, Ming Ding & Longgang Tian
School of Electrical Engineering and Automation, Hefei University of Technology, Hefei, China

Xuesong Zhang & Xiaohui Ge
Z(P)EPC Electric Power Research Institute, Hangzhou, China

ABSTRACT: With the development of micro-grid technology, hybrid AC/DC micro-grid owns both AC and DC micro-grid advantages, and fully considers the characteristics of distributed power, energy storage, and customer load, so now it has become the research focus. According to the research on AC and DC micro-grid at home and abroad, we analyze and summarize the research methods of hybrid AC/DC micro-grid, including its control strategy and stability analysis. In order to study the stability of hybrid AC/DC micro-grid in the grid-connected and stand-alone operation mode, this paper discusses hybrid AC/DC micro-grid stability analysis method on different cases and proposes hybrid AC/DC micro-grid stability factors, providing the appropriate reference to the subsequent research on hybrid AC/DC micro-grid.

Keywords: hybrid AC/DC micro-grid; control strategy; small-signal stability

1 INTRODUCTION

With the conventional energy sources gradually exhausting and environmental pollution becoming worse, the countries all over the world begin to focus on environmentally friendly, efficient, and flexible power generation—Distributed Generation. In order to improve the utilization of renewable energy sources, researchers have proposed a new distribution method—micro-grid. It earns the features of small environmental pollution: high reliability of power supply, digesting the local power, and micro-grid has developed rapidly. Most countries research on AC micro-grid and DC micro-grid, and then it comes to hybrid AC/DC micro-grid. AC micro-grid is still the main form of micro-grid. Although there are many research results of AC micro-grid, it still needs to be further addressed in resonance and harmonic aspects caused by distributed power Parallel accessing; DC micro-grid does not need to focus on synchronization problems between the various DG, and has more advantages in circulation suppression; nevertheless, the DC load type and capacity limit its development.[1,4] Hybrid AC/DC micro-grid can reduce power loss and harmonic currents caused by a couple of AC/DC or DC/AC conversion, which applied to the AC or DC micro-grid. What is more, it can also improve the reliability and economy of micro-grid system, and makes full use of a variety of renewable power sources, energy storage devices, and loads. So the hybrid AC and DC micro-grid won the attention of research scholars in the word, and it will earn a rapid development.

Nowadays, the research on hybrid AC/DC micro-grid is just starting the whole world, and there has not built a complete research system of hybrid AC/DC micro-grid's grid structure, control strategy, and stability. For the small-signal stability for hybrid AC/DC micro-grid, researchers have not focused much on it. Hybrid AC/DC micro-grid uses different control strategies because of its different grid structure, so the small-signal stability methods are also not the same. This paper discusses hybrid AC/DC micro-grid stability in different operation modes.

2 HYBRID AC/DC MICRO-GRID BASIC GRID STRUCTURE

A hybrid AC/DC micro-grid is composed of AC micro-grid system and DC micro-grid system; it has both AC and DC bus, so it can directly provide power to AC and DC load. A hybrid AC/DC micro-grid contains a group of power electronic devices; therefore, it needs to reasonably design the grid structure of the micro-grid in order to optimize distributed generations, energy storage device, and load. A reasonable topological structure can improve the reliability and flexibility of micro-grid system when connecting to low-voltage distribution network; when it comes to design voltage level of hybrid AC/DC micro-grid, it needs to meet the characteristics of renewable energy and

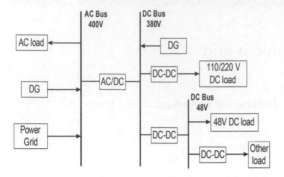

Figure 1. A typical hybrid AC/DC micro-grid structure.

load, then it will reduce power electronic converter and also can improve system stability; for different distributed energies inserting into micro-grid, we should consider all kinds of factors, location, capacity, output voltage level, and so on, to choose a suitable connecting methods.[6]

There are various structures of hybrid AC/DC micro-grid because power grid, AC micro-grid system, and DC micro-grid system have different patterns to be connected. Power grid provides power for micro-grid load to keep voltage level and frequency stable; DC micro-grid system can directly connect to power grid, then use AC/DC converter to join AC micro-grid system into the whole system; and also can connect AC micro-grid system to power grid, then link DC micro-grid system into AC bus by converter.

Combining the key scientific and technological project of Zhejiang Electric Power Corporation, "Low-voltage Intelligent Hybrid AC/DC Micro-grid Key Technology Research and Demonstration Project Construction", this paper designs a typical hybrid AC/DC micro-grid grid structure, as shown in Figure 1.

Where, 400 V AC bus directly connects to power grid, at the same time, link a part of new energy and AC load to AC bus; insert an AC/DC converter between AC and DC micro-grid in order to achieve AC and DC system hybrid; DC micro-grid use double-bus structure, directly providing power to DC load. This hybrid AC/DC micro-grid structure completely meets the basic design principles of micro-grid system.

3 SMALL-SIGNAL STABILITY ANALYSIS FOR MICRO-GRID

Small-signal stability for power system means the ability to keep the system synchronous after small disturbance. It depends on the initial operation state, and the characteristics of various control devices. There are many methods to analyze small-signal stability for power system. According to the mathematical model, it includes eigenvalue analysis, digital simulation method, and frequency domain analysis. Eigenvalue analysis has already been one of the most effective methods in power system dynamic stability analysis.[2,3]

3.1 Analysis of small-signal stability

Power system could be described as n first-order nonlinear ordinary differential algebraic equations:

$$\dot{x} = f(x, k) \tag{1}$$

The output variables are defined as:

$$y = g(x, k) \tag{2}$$

where n is the dimension of the system, k the input variable of the system, x the state variable, y the output variable, and g a nonlinear function vector of the combination of input and output variables.

In the analysis of power system small-signal stability, disturbances produced by load and generators are usually small enough to ignoring them; the system nonlinear equations could be linearized on the initial stable operation point, then get the approximate linear equation state.

Eq.1 and Eq.2 could be linearized to:

$$\begin{cases} \Delta\dot{x} = A\Delta x + B\Delta k \\ \Delta y = C\Delta x + D\Delta k \end{cases} \tag{3}$$

where A is the partial derivative of function f(x, k):

$$A = \begin{bmatrix} \dfrac{\partial f_1}{\partial x_1} & \cdots & \dfrac{\partial f_1}{\partial x_n} \\ \vdots & \ddots & \vdots \\ \dfrac{\partial f_n}{\partial x_1} & \cdots & \dfrac{\partial f_n}{\partial x_n} \end{bmatrix} \tag{4}$$

Stability of nonlinear system depends on the eigenvalues of state matrix A. According to the first law of Lyapunov, system would be stable when all the eigenvalues of state matrix A have negative real parts, and in other conditions the system would be unstable.

3.2 Small-signal Stability Analysis of Micro-grid

With the increasing penetration of distributed energy, the impacts on grid and the complexity between micro-grid and grid cannot be ignored anymore. A huge group of power electronics devices are utilized in micro-grid, but reduced micro-grid system inertial with small inverter output reactance and fast response of power electronics devices, and increase the challenge on analysis of system stability. In small-signal analysis for micro-grid, network-load small-signal model and inverter small-signal model should all be considered, in the modeling of complete micro-grid small signal model. With regard to hybrid AC-DC micro-grid, there are multiple operation states. We will consider several different hybrid AC-DC micro-grid modes for small-signal analysis in the following parts.

4 SMALL-SIGNAL ANALYSIS FOR HYBRID AC-DC MICRO-GRID IN GRID-CONNECTED OPERATION MODE

When hybrid AC-DC micro-grid is in grid-connected operation mode, load is mainly energized by the power grid, and frequency and voltage level are also kept in normal region with power gird. AC- and DC-distributed generators just take part in energy exchanging.

4.1 Network-load small-signal model

Integrated in external grid, the modeling of network-load small-signal model should be considered connected to infinity source and the DC part could be modeled as active AC load connected to AC bus. In the grid-connected operation mode, the equivalent circuit of hybrid AC-DC micro-grid is shown in Figure 2.

Where, R_n and L_n are equivalent link reactance between external grid and micro-grid, R_1 and L_1

are line impedance, and R_{load} and L_{load} are load impedance; DC micro-system is equivalent to the link circuit between AC generators and equivalent impedance.

According to Figure 2, network-load state equations and node voltage equations would be deduced; with Park transformation and linearization, network-load small-signal model can be described.

4.2 Inverter small-signal model

As for the hybrid AC-DC micro-grid mode in grid-connected operation mode, DC/DC and DC/AC converters are controlled in different strategies. Converter averaged model would be utilized for easily building small-signal model. Averaged model ignores fast switching state, utilizes equivalent electrical functions to represent inverter characteristics, and simplifies the process of inverter small-signal modeling. Inverter small-signal model also contains inverter interface small-signal model, control small-signal model, and so on, then integrate them into small-signal model.[8,11]

For a complete hybrid AC/DC micro-grid small-signal model, it needs to combine network-load small-signal model and inverter small-signal model, then it comes to system small-signal model:

$$[\Delta \dot{x}_n] = A_n[\Delta x_n] \qquad (5)$$

where A_n is system characteristic matrix. Analysis its characteristic value by MATLAB, and research the stability of hybrid AC/DC micro-grid system and factors influencing system stability.

5 SMALL-SIGNAL STABILITY ANALYSIS FOR HYBRID AC/DC MICRO-GRID IN STAND-ALONE OPERATION MODE

When hybrid AC/DC micro-grid is in stand-alone operation mode, there are two situations: 1. AC micro-grid system is connected to the power grid, but DC micro-grid is stand-alone. At this time, the PCC (Point of Common Coupling) node of bidirectional AC/DC inverter is breaking, so the distributed energy in DC micro-grid system provides the power to DC load in this system, to keep system voltage and frequency stable. As shown in Figure 3, it means that PCC-1 node is closing and PCC-2 node is breaking. 2. The PCC node, which connects hybrid AC/DC micro-grid system and the power grid, is breaking, then AC micro-grid and DC micro-grid are both in stand-alone operation mode. At this time, distributed energy of the whole micro-grid system provides power to AC and DC

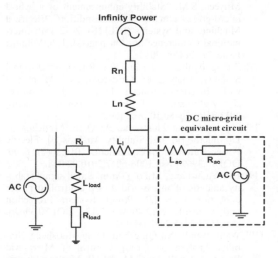

Figure 2. Hybrid AC-DC micro-grid equivalent circuit.

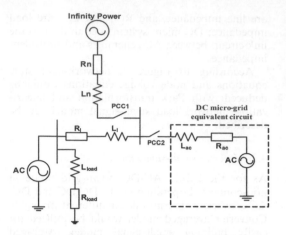

Figure 3. Hybrid AC-DC micro-grid equivalent circuit in stand-alone operation mode.

load, and it keeps the system stable. In Figure 3, it means that PCC-1 node is breaking and PCC-2 node is closing.

When hybrid AC-DC micro-grid system is in the NO.1 stand-alone operation mode, the stability analysis method for AC micro-grid system, which is connected to power grid, is similar to it in grid-connected operation mode, that is ignoring equivalent circuit and the converter small-signal model of DC micro-grid system; for the DC micro-grid system in stand-alone operation mode, its small-signal analysis can use converter average model, and then research its characteristic value. When hybrid AC-DC micro-grid system is in the NO.1 stand-alone operation mode, the bi-directional AC/DC inverter, which connects AC and DC micro-grid system, becomes the important device of the micro-grid system, and it can keep the system stability by its control. Built converter small-signal model based on converter control-circuit, then built the system small-signal model in stand-alone operation mode by combining the converter small-signal model and network-load small-signal model, analyze the model to judge the system stable or unstable.

6 SUMMARY

The small-signal stability analysis for a hybrid AC/DC micro-grid system, which contains a huge group of power electronics devices, needs to comprehensively consider network, load, converter (including its control-circuit and harmonic treatment circuit) small-signal model; at the same time, when the micro-grid system is in different operation modes, it needs to depend on the actual situation

to discuss and analyze. There are many other factors, which influence the stability of hybrid AC/DC micro-grid system, such as line impedance, equivalent link impedance, and converter control parameters. When analyzing the micro-grid stability, it needs to consider all of these factors, in order to build a stable micro-grid system.

ACKNOWLEDGEMENTS

Thanks for State Grid Zhejiang Electric Power Company's project. Project NO.5211011306Z6.

REFERENCES

[1] Huang Wen-tao, Tai Neng-ling, Fan Chun-ju, Lan Sen-lin, Tang Yue-zhong, Zhong Yong. Study on structure characteristics and designing of micro-grid [J]. Power System Protection and Control, 2012,18(40):149–155.

[2] Suning. Small-Signal Stability Analysis of Micro-grid Control and Strategy for Micro-grid Energy Management [D]. North China Electric Power University (Beijing), 2011. DOI:10.7666/d.y1954945.

[3] Xiao Zhaoxia, Wang Chengshan, Wang Shouxiang. Small-signal Stability Analysis of Micro-grid Containing Multiple Micro Sources [J]. Automation of Electric Power Systems, 2009, 33(6):81–85. DOI:10.3321/j.issn:1000-1026.2009.06.019.

[4] I.J. Balaguer, Q. Lei, S. Yang, U. Supatti, and F.Z. Peng, "Control for grid-connected and intentional islanding operations of distributed power generation," IEEE Trans. Ind. Electron., vol. 58, no. 1, pp. 147–157, January 2011.

[5] X. Liu, P. Wang, and P.C. Loh, "A hybrid AC/DC microgrid and its coordination control," IEEE Trans. Smart Grid, vol. 2, no. 2, pp. 278–286, January 2011.

[6] Ambia, M.N. Al-Durra, A. Caruana, C. Muyeen, S.M. "Stability enhancement of a hybrid micro-grid system in grid fault condition", Electrical Machines and Systems (ICEMS), 2012 15th International Conference on, On page(s): 1–6, Volume: Issue: 21–24 Oct. 2012.

[7] Zhang Jianhua, Su Ling, Liu Ruoxi. Small-signal Stability Analysis of Grid-connected Micro-grid with Inverter-interfaced Distributed Resources [J]. Automation of Electric Power Systems, 2011,35(6):76–80,102.

[8] Shi Jie, Zheng Zhanghua, Ai Qian. Modeling of DC micro-grid and Stability Analysis [J]. Electric Power Automation Equipment, 2010,30(2):86–90. DOI:10.3969/j.issn.1006-6047.2010.02.018.

[9] Fan Yuan-liang, Miao Yi-qun. Small signal stability analysis of micro-grid droop controlled power allocation loop [J]. Power System Protection and Control, 2012,40(4):1–7,13. DOI:10.3969/j.issn.1674-3415.2012.04.001.

[10] Wang Wendi, Xu Qingshan, Ding Maosheng, Stability Analysis on Droop Control of Micro-grid Based on Small-signal Model [J]. Modern Electric Power, 2014,31(3):17–21.

[11] Mao Meiqin, Ding Ming, Chang Liuchen. Tested and Information Integration of EMS for a Micro-grid with Multi-energy Generation Systems [J]. Automation of Electric Power Ssystems, 2010,34(1):106–111.

[12] Yin Xiao-gang, Dai Dong-yun, HAN Yun. Discussion on Key Technologies of AC-DC Hybrid Microgrid [J]. High Voltage Apparatus, 2012,48(9):43–46.

[13] Wang Chengshan, Sun Chongbo, Peng Ke. Study on AC-DC Hybrid Power Flow Algorithm for Microgrid [J]. Proceedings of the CSEE, 2013,33(4):8–15.

[14] Li Cong. Operation Simulation and Small-signal Stability Analysis of Micro-grid Based on Droop Control [D]. Southwest Jiaotong University, 2013. DOI:10.7666/d.Y2334993.

[15] Wu Weimin, He Yuanbin, Geng Pan, Qian Zhaoming, Wang Yousheng. Key Technologies for Micro-Grids [J]. Transactions of China Electrotechnical Society, 2012,1(27):98–113.

[16] Dos Santos, E.C. Alibeik, M. Creek, B. New power electronics converter interfacing a DG system with hybrid dc/ac micro-grid [J]. 2013 IEEE Power and Energy Conference at Illinois: 180–185.

Electrical, Control Engineering and Computer Science – Liu (Ed.)
© *2016 Taylor & Francis Group, London, ISBN 978-1-138-02937-8*

Plasma slab's electromagnetic behavior calculation based on two Auxiliary Differential Equation (ADE) styles

Jinzu Ji, Hongqian Wu & Peilin Huang
School of Aeronautic Science and Engineering, Beihang University, Beijing, China

ABSTRACT: Two styles of Auxiliary Differential Equation (ADE) were applied and compared in this paper to study the electromagnetic behavior in plasma slab. One ADE samples polarization current at integer time step and the other samples it at half integer time step. In implication via computer code, the former method need to store one more back level electric field to update polarization current, while the other need not. Numerical results are compared with exact and Piecewise Linear Recursive Convolution (PLRC) solution to validate the algorithm. The results show that the two ADE styles have almost the same accuracy as PLRC.

Keywords: plasma; Finite-Difference Time-Domain (FDTD); Auxiliary Differential Equation (ADE); reflection; transmission

1 INTRODUCTION

The Finite-Difference Time-Domain (FDTD) has become an important methodology in manipulating electromagnetic problems since Yee raised his famous Yee's leapfrog grid scheme [1]. It discretizes the Maxwell's equations directly and can get spectral characteristics by one-time calculation because it is a time-domain method. Originally, the scheme was created to discretize Maxwell's equations with the assumption that the medium that hosted in the electromagnetic wave was isotropic, non-dispersive, linear and time-invariant. But, in real world, many media have strong dispersive performance, such as earth, water, plasma, and so on. As the method grew in popularity and proved its worthiness in the prediction of scattering, diffraction and propagation events, many researchers postulated and then devised ways that the algorithm could be applied to problems for which the medium was dispersive. These algorithms includes Recursive Convolution (RC) [2, 3], frequency-dependent Z Transform (ZT) [4], Piecewise Linear Recursive Convolution (PLRC) [5], auxiliary differential equation [6, 7], Trapezoidal Recursive Convolution (TRC) [8], and JE Convolution (JEC) [9]. Although PLRC often attains higher accuracy than RC, it requires complicated formulation of two convolution integrals and one more back level electric field. The TRC method requires single convolution integral in the formulation as in RC, while maintaining the accuracy comparable to the PLRC.

This paper examines two ADE styles in calculating electromagnetic behavior in dispersive media.

A cold plasma slab's reflection and transmission coefficients are calculated and validated with the accurate results. To demonstrate the advantage of the ADE method, Piecewise Linear Recursive Convolution (PLRC) results are also presented and compared.

Time factor $e^{j\omega t}$ is assumed and suppressed throughout this paper.

2 FORMULATION

Ampere's law in Maxwell's equations can be expressed in the form

$$\nabla \times \mathbf{H} = j\omega\varepsilon_0\varepsilon_r\mathbf{E} \tag{1}$$

where $\mathbf{H}$ and $\mathbf{E}$ are magnetic and electric intensities, respectively, ε_0 is permittivity in free space, ε_r is relative permittivity. ε_r can be expressed as sum of constant and dispersive parts. For cold plasma medium, ε_r can be expressed in form

$$\varepsilon_r = 1 + \omega_p^2 / \left(j\omega\left(j\omega + v_c\right)\right) \tag{2}$$

where ω_p is plasma's frequency, ω is incident electromagnetic wave's frequency, v_c is plasma's collision frequency with neutral particles. To emphasize polarization current $\mathbf{J}_p$. The Ampere's law can be represented in form

$$\nabla \times \mathbf{H} = j\omega\varepsilon_0\mathbf{E} + \mathbf{J}_p \tag{3}$$

where $\mathbf{J}_p$ is defined as

$$J_p = \omega_p^2/(j\omega + \nu_c)\mathbf{E}. \tag{4}$$

The time-domain expression of (4) is

$$\frac{\partial \mathbf{J}_p}{\partial t} + \nu_c\mathbf{J}_p = \omega_p^2\mathbf{E}. \tag{5}$$

By the convention of Yee's staggered grids, electric field is sampled at integer time steps and magnetic field is sampled at half integer time steps. There are two styles for polarization current's sampling styles that one is sampling at integer time steps and the other is sampling at half integer time steps.

If $\mathbf{J}_p$ is sampled at integer time steps, (3) can be discretized as

$$\nabla \times \mathbf{H}^{n+1/2} = \varepsilon_0\left(\mathbf{E}^{n+1} - \mathbf{E}^n\right)\Big/\Delta t + \left(\mathbf{J}_p^{n+1} + \mathbf{J}_p^n\right)\Big/2 \tag{6}$$

and (5) can be discretized as

$$\frac{\mathbf{J}_p^{n+1} - \mathbf{J}_p^n}{\Delta t} + \frac{\nu_c}{2}\left(\mathbf{J}_p^{n+1} + \mathbf{J}_p^n\right) = \frac{\omega_p^2}{2}\left(\mathbf{E}^{n+1} + \mathbf{E}^n\right) \tag{7}$$

In (6) and (7), $\mathbf{J}_p$ and $\mathbf{E}$ are all time averaged at time step $n + 1/2$. (7) can be expressed in explicit form

$$\mathbf{J}_p^{n+1} = k_p\mathbf{J}_p^n + \beta_p\left(\mathbf{E}^{n+1} + \mathbf{E}^n\right) \tag{8}$$

where

$$k_p = \left(2 - \nu_c\Delta t\right)/\left(2 + \nu_c\Delta t\right)$$
$$\beta_p = \Delta t\varepsilon_0\,\omega_p^2\Big/\left(2 + \nu_c\Delta t\right). \tag{9}$$

Substituting (8) into (6) and expressing $\mathbf{E}^{n+1}$ in explicit form, we get

$$\mathbf{E}^{n+1} = \mathrm{CA}\cdot\mathbf{E}^n - \mathrm{CB}\cdot\frac{1 + k_p}{2}\mathbf{J}_p^n + \mathrm{CB}\cdot\nabla\times\mathbf{H}^{n+1/2} \tag{10}$$

where

$$\mathrm{CA} = (2\varepsilon_0 - \beta_p\Delta t)/(2\varepsilon_0 + \beta_p\Delta t)$$
$$\mathrm{CB} = 2\Delta t/(2\varepsilon_0 + \beta_p\Delta t). \tag{11}$$

From (8) and (10), $\mathbf{E}^n$ and $\mathbf{J}_p^n$ can be achieved by iteration. In calculation of $\mathbf{J}_p^{n+1}$, both $\mathbf{E}^{n+1}$ and $\mathbf{E}^n$ are needed, thus one more back level of electric field be stored.

If we sample $\mathbf{J}_p$ at half integer time steps, we will get a different iterative form. Equation (3) can be discretized as

$$\nabla \times \mathbf{H}^{n+1/2} = \varepsilon_0\left(\mathbf{E}^{n+1} - \mathbf{E}^n\right)\Big/\Delta t + \mathbf{J}_p^{n+1/2} \tag{12}$$

and the explicit form is

$$\mathbf{E}^{n+1} = \mathbf{E}^n - \Delta t/\varepsilon_0\,\mathbf{J}_p^{n+1/2} + \Delta t/\varepsilon_0\,\nabla\times\mathbf{H}^{n+1/2} \tag{13}$$

(5) can be discretized as

$$\left(\mathbf{J}_p^{n+1/2} - \mathbf{J}_p^{n-1/2}\right)\Big/\Delta t + \nu_c\left(\mathbf{J}_p^{n+1/2} + \mathbf{J}_p^{n-1/2}\right)\Big/2 = \omega_p^2\mathbf{E}^n. \tag{14}$$

Equation (14)'s explicit form is

$$\mathbf{J}_p^{n+1/2} = k_p\mathbf{J}_p^{n-1/2} + 2\beta_p\mathbf{E}^n \tag{15}$$

where k_p and β_p are expressed in (9). Equations (13) and (15) are the iterative form to update $\mathbf{E}^n$ and $\mathbf{J}_p^n$. The advantage of this style is that there is no need to store back level electric field to update polarization current.

3 NUMERICAL VALIDATION

Calculations were made for a plasma slab 15 mm thick to validate the two ADE styles. The one-dimensional problem space consists of 500 spatial cells with 75 μm thickness with the plasma slab occupying cells 200 through 400. The time step is 0.125 ps. The plasma considered has a plasma frequency of 28.7 GHz ($\omega_p/2\pi$) and a collision frequency ν_c of 20 rad/s with neutral particles.

Calculations were conducted for a Gaussian pulse normally incident plane wave. The spectrum of the incident pulse rises sharply but smoothly from zero frequency, peaks at approximately 50 GHz. In order to establish the accuracy of the results, a comparison with the exact results for the reflection and transmission coefficients is made by Fourier Transformation (FT) of the ADE time-domain results.

The reflection and transmission coefficients were obtained from the ADE results by calculating the electric field strength versus time in the spatial cells just in front of (cell number 200) and behind (cell number 400) the plasma slab. The transmitted field versus time is just the field in cell 400, while the reflected field is the subtraction of total field and incident field in cell 200. The complex reflection or transmission coefficients at each frequency were calculated by dividing the frequency domain forms of the reflection or transmission and incident fields.

Totally 6000 time steps were included in iteration. The exact analytical results for steady state complex reflection and transmission coefficients versus frequency were obtained using the complex permittivity. Figure 1 shows the reflection and transmission coefficients magnitude in dB, respectively. The calculation results agree well with exact solution. To demonstrate the accuracy compared to other methods manipulating dispersive medium, PLRC results are also illustrated in Figure 1. The comparison shows that ADE has equal accuracy with PLRC method.

In Figure 2, the phases of reflection and transmission coefficients are illustrated. The Figures show that ADE can also calculate the phase accurately.

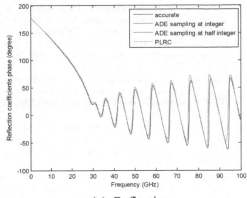

(a) Reflection

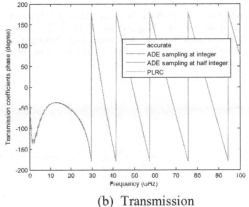

(b) Transmission

Figure 2. Reflection and transmission coefficient phase.

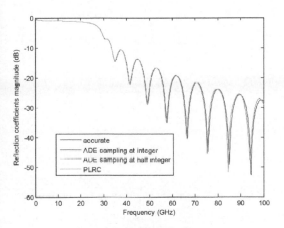

(a) Reflection

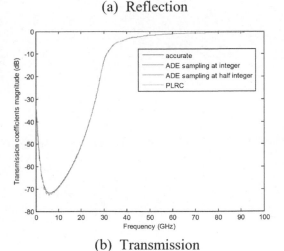

(b) Transmission

Figure 1. Reflection and transmission coefficient magnitude.

4 CONCLUSION

Reflection and transmission coefficients of plasma slab are calculated in the time-domain with two ADE styles. Polarization current samples at different time steps in the two ADE styles. If it samples at integer time steps, one more back level electric field should be stored. While it samples at half integer time steps, extra back level field storage is avoided. The numerical results validate the accuracy of both ADE styles and show that they have almost equal accuracy with PLRC.

REFERENCES

[1] Yee, K.S., 'Numerical Solution of Initial Boundary Value Problems Involving Maxwell's Equations in Isotropic Media', Antennas and Propagation, IEEE Transactions on, 1966, 14, (3), pp. 302–307, doi: 10.1109/TAP.1966.1138693.

[2] Luebbers, R., Hunsberger, F.P., Kunz, K.S., Standler, R.B., and Schneider, M., 'A Frequency-Dependent Finite-Difference Time-Domain Formulation for Dispersive Materials', Electromagnetic Compatibility, IEEE Transactions on, 1990, 32, (3), pp. 222–227, doi: 10.1109/15.57116.

[3] Luebbers, R.J. and Hunsberger, F., 'Fdtd for Nth-Order Dispersive Media', Antennas and Propagation, IEEE Transactions on, 1992, 40, (11), pp. 1297–1301, doi: 10.1109/8.202707.

[4] Sullivan, D.M., 'Frequency-Dependent Fdtd Methods Using Z Transforms', Antennas and Propagation, IEEE Transactions on, 1992, 40, (10), pp. 1223–1230, doi: 10.1109/8.182455.

[5] Kelley, D.F. and Luebbers, R.J., 'Piecewise Linear Recursive Convolution for Dispersive Media Using Fdtd', Antennas and Propagation, IEEE Transactions on, 1996, 44, (6), pp. 792–797, doi: 10.1109/8.509882.

[6] Kashiwa, T., Yoshida, N., and Fukai, I., 'Transient Analysis of a Magnetized Plasma in Three-Dimensional Space', Antennas and Propagation, IEEE Transactions on, 1988, 36, (8), pp. 1096–1105, doi: 10.1109/8.7222.

[7] Gandhi, O.P., Gao, B.Q., and Chen, J.-Y., 'A Frequency-Dependent Finite-Difference Time-Domain Formulation for General Dispersive Media', Microwave Theory and Techniques, IEEE Transactions on, 1993, 41, (4), pp. 658–665, doi: 10.1109/22.231661.

[8] Siushansian, R. and LoVetri, J., 'A Comparison of Numerical Techniques for Modeling Electromagnetic Dispersive Media', Microwave and Guided Wave Letters, IEEE, 1995, 5, (12), pp. 426–428, doi: 10.1109/75.481849.

[9] Qing, C., Katsurai, M., and Aoyagi, P.H., 'An Fdtd Formulation for Dispersive Media Using a Current Density', Antennas and Propagation, IEEE Transactions on, 1998, 46, (11), pp. 1739–1746, doi: 10.1109/8.736632.

Electrical, Control Engineering and Computer Science – Liu (Ed.)
© 2016 Taylor & Francis Group, London, ISBN 978-1-138-02937-8

The electromagnetic equivalent modeling and simulation analysis of large size metal structure

Penghao Xie
*State Key Laboratory of Control and Simulation of Power Systems and Generation Equipment
(Department of Electrical Engineering, Tsinghua University), Beijing, China
Institute of Electrostatic and Electromagnetic Protection, Mechanical Engineering College, Shijiazhuang, China*

Jiansheng Yuan
*State Key Laboratory of Control and Simulation of Power Systems and Generation Equipment (Department of
Electrical Engineering, Tsinghua University), Beijing, China*

Junjian Bi
Institute of Electrostatic and Electromagnetic Protection, Mechanical Engineering College, Shijiazhuang, China

ABSTRACT: In order to study the electromagnetic coupling law of one typical metal body, electromagnetic simulation software FEKO is utilized to set up the typical metal body model. With the increase of the simulation frequency, different lengths of metal body internal coupling field curve gradually separated, has certain regularity of fitting. Preliminary determination by reducing the length of the metal equivalent for the test model is established and confirmed in law of fitting better experimental spectrum.

Keywords: simulation; metal body; electromagnetic coupling

1 INTRODUCTION

Abroad mostly entity modeling and simulation is used for the electronic equipment of metal packaging, and most domestic research on seam hole electromagnetic coupling rule is aimed at small size metal shell, using the finite difference time domain method for a single simple hole seam model simulation, considering less factors such as electronic components inside the equipment. For the problems of larger size metal structures in the laboratory test area to carry out the irradiation test of control unit and other components, research on the electromagnetic equivalent transform to the length of the metal structure is needed.

2 THE PHYSICAL MODELING OF TYPICAL METAL STRUCTURE

2.1 *The choice of electromagnetic simulation software*

Software FEKO is based on the method of moments, adopting multilevel fast multi-stage algorithm greatly improves the computational efficiency on the premise of keeping precision. The method of moments combined with high frequency analysis method are used to solve the Radar scattering Cross Section (RCS), open field radiation, electromagnetic compatibility of the various electromagnetic field analysis problems. This selection of electromagnetic simulation software FEKO for simulation calculation is more appropriate.

2.2 *The choice of frequency points and the metal body external electromagnetic environment*

On the choice of experimental frequency simulation environment and the external side, the standard GJB xxxx is used as reference. The unit of electric field intensity in the standard is V/m, but using dBV/m is more appropriate in analysis process for the simulation calculation results, in order to facilitate the calculation of shielding effectiveness of metal body. A0 card in FEKO is used to set up linear polarization plane electromagnetic wave field, as shown in Figure 1. In A0, the "New source" and "Linear" are selected, the electric field strength is set as 50V/m, θ and φ are 0°, along the x axis. The frequency range is from 100 MHz to 600 MHz, and the step length is 100 MHz.

2.3 *The establishment of physical model*

For solid modeling, the size of a certain type of metal body was measured, and metal structure is

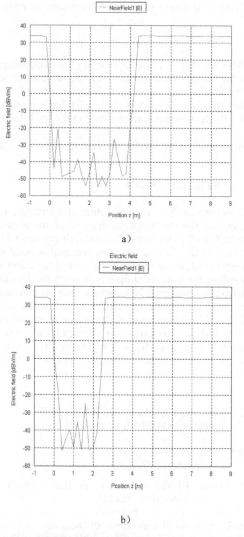

Figure 1. A0 card.

simplified refer to related information as shown in Figure 2. For the metal body is axisymmetric model, the bottom of the metal body center as the origin of coordinates, the center of the metal body axis is established as the z axis. Metal body (including the control unit) overall length is xx mm, and metal body main material is the ideal electrical conductor.

3 THE ANALYSIS OF SIMULATION RESULTS

Control unit length is constant, the metal body length not including the control unit are L and 0.5 L respectively, and the corresponding

Electric field

a)

Electric field

b)

Figure 3. x = 0 y = 0 100 MHz.

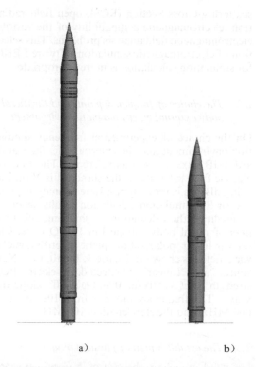

a)   b)

Figure 2. The metal body diagram.

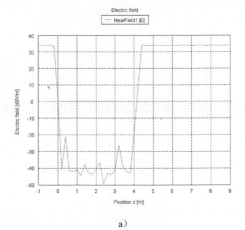

a)

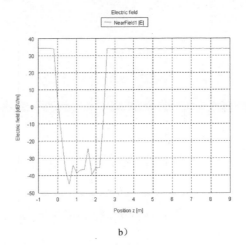

b)

Figure 5.   x = 0 y = 0 300 MHz.

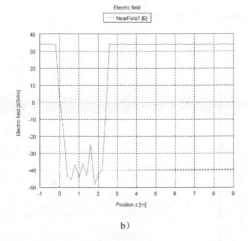

b)

Figure 4.   x = 0 y = 0 200 MHz.

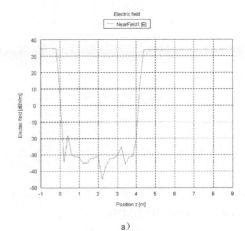

a)

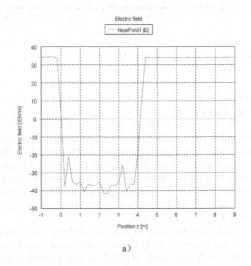

a)

Figure 5.   (*Continued*)

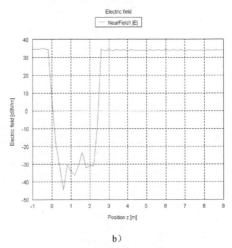

b)

Figure 6.   x = 0 y = 0 400 MHz.

65

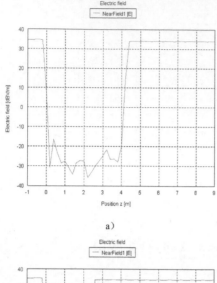

a)

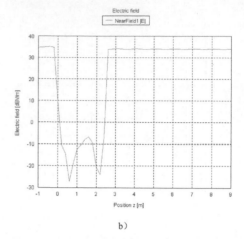

b)

Figure 8.   x = 0 y = 0 600 MHz.

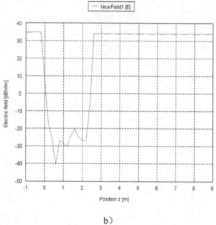

b)

Figure 7.   x = 0 y = 0 500 MHz.

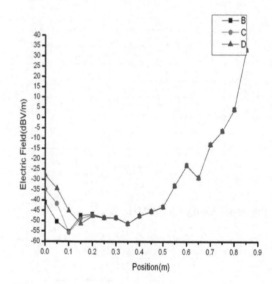

Figure 9.   x = 0 y = 0 100 MHz.

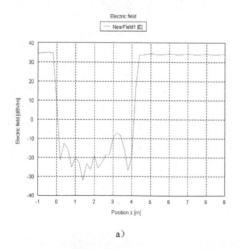

a)

Figure 8.   (*Continued*)

simulation results are in the figure a), b) (seen from Fig. 3 to Fig. 8).

In order to analyze the electromagnetic coupling law of control unit, setting the control unit as simulation object, the metal body excluding the control unit of length L and 0.5 L respectively two cases belong to the same coordinate system, simulation results corresponding to the following figure the curve C, D. Curve B is calibration curve (seen from Fig. 9 to Fig. 14).

With the increase of the simulation frequency, internal coupling field curve in different lengths of metal body gradually separated, has certain regularity of fitting. Internal coupling field curve of

66

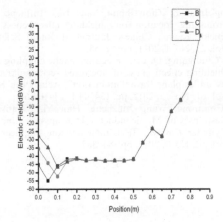

Figure 10. x = 0 y = 0 200 MHz.

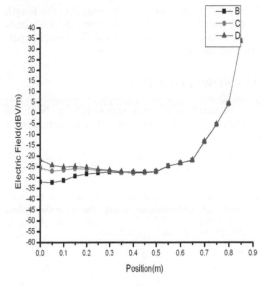

Figure 13. x = 0 y = 0 500 MHz.

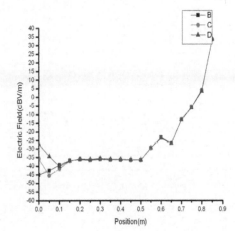

Figure 11. x = 0 y = 0 300 MHz.

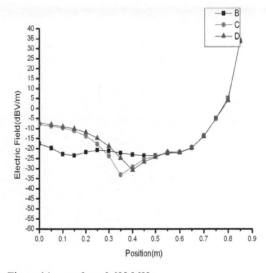

Figure 14. x = 0 y = 0 600 MHz.

the control unit with different lengths of metal has good fitting.

4 CONCLUSIONS

With the increase of the simulation frequency, internal coupling field curve in different lengths of metal body has certain regularity of fitting. Internal coupling field curve of the control unit with different lengths of metal has good fitting.

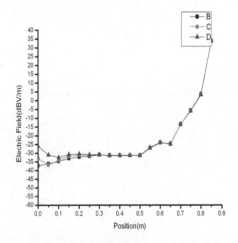

Figure 12. x = 0 y = 0 400 MHz.

Preliminary determination by reducing the length of the metal equivalent for the test model is established and confirmed in law of fitting better experimental spectrum.

ACKNOWLEDGMENT

The research work was supported by National Natural Science Foundation of China under Grant No. 51207168.

REFERENCES

Tesche F.M, "On the analysis of a transmission line with nonlinear terminations using the time-dependent BLT equation", IEEE Trans. Electromagn. Compat, Vol. 49, No. 2, 2007, pp. 427–433.

Wang Jianbao, Zhou Bihua, Yang Bo, "Influence of grounding location on shielding effectiveness of shielde room", Chinese Journal of Radio Science, Vol. 27, No. 1, 2011, pp. 52–55.

Jiao Chongqing, Qi Lei, "Electromagnetic coupling and shielding effectiveness of apertured rectangular cavity under plane wave illumination", Acta. Phys. Sin., Vol. 61, No. 13, 2012, pp. 134104.1–134104.6.

Li Chunrong, Wang Xinzheng, Han Yi, "Transient Responses of Multiconductor Transmission Lines in Cylinder Cavity", Telecommunication Engineering, Vol. 52, No. 3, 2012, pp.384–387.

Electrical, Control Engineering and Computer Science – Liu (Ed.)
© 2016 Taylor & Francis Group, London, ISBN 978-1-138-02937-8

Transmission characteristics of Quadruple-Ridge Square Waveguide loaded with right-handed and left-handed materials

Guojian Li & Yinqin Cheng
College of Electrical Engineering, Northwest University for Nationalities, Lanzhou, China

Aning Ma
School of Information Science and Engineering, Lanzhou University, Lanzhou, China

ABSTRACT: This paper presents a numerical study of the dielectric loaded Quadruple-Ridged Square Waveguide (QRSW). The dispersion characteristics of left-handed and right-handed material-loaded QRSW are first investigated by the edge element method. The variations of the cut-off wavelength and single-mode bandwidth with the ridge dimensions for different values of dielectric constant are studied in detail. The numerical results in this paper provide an extension to the design for the ridge waveguide and are helpful in practical applications for millimeter waves.

Keywords: quadruple-ridged square waveguide; finite element method; dispersion characteristics

1 INTRODUCTION

Original publication on ridge waveguide dates back to 1947 [1]. Since that time, ridge waveguide and its variations have been widely studied in conjunction with microwave and millimeter wave applications, because of their unique characteristics such as low cut-off frequency, wide bandwidth, and low impedance characteristics [2].

Ridge waveguide partially filled with dielectric have an important feature, i.e., ridge waveguide can be miniaturized by loading dielectric in it, which is very useful in practical applications [3–5].

In the past decade, tremendous interests have occurred with the experimental realization of the possibility of creating novel Left-Handed Materials (LHM), which do have simultaneously negative effective and permittivity and effective and permeability. Artificial negative magnetic permeability was proposed by Pentry et al. [6], and experimentally demonstrated by Smith et al. by creating novel type of microstructure, such as array of metallic split-resonator [7]. Krowne et al. have systematically studied the electromagnetic behavior in LHM loaded microstrip structures [8,9].

This article will discuss the transmission characteristics, such as the cut-off wavelength and single-mode bandwidth in dielectric loaded Quadruple-Ridge Square Waveguide for millimeter waves (QRSW) by employing the edge-based FEM. By ensuring only tangential continuity of the field components across element boundary completely the edge-based FEM wipe out the spurious modes.

2 THEORETICAL ANALYSES

To analyze electromagnetic field in an inhomogeneous waveguide, the edge-based finite element method is employed in the framework of the Galerkin formulation of the weighted residual method to solve the vector Helmholtz equations. Spurious modes are avoided using mixed edge and nodal elements to approximate the transverse and longitudinal components of electric field, respectively.

The QRSW is shown in Figure 1. a is the outer dimensions of the square waveguide. b and d are used to describe the dimensions of the rectangular ridge. ε_r is the relative permittivity of free space.

Owing to the dielectric loaded, electromagnetic waves propagating in QRSW are neither pure TE mode nor pure TM mode, but a combination of them, both $\mathbf{E}_z$ and $\mathbf{H}_z$ exist simultaneously. Modes in QRSW may in general be described as quasi-LSE or quasi-LSM ones depending on whether $\mathbf{E}_x$ or $\mathbf{E}_y$ is equal to zero. Maxwell's equations in QRSW can be written as

$$\nabla_T \times \mathbf{H} = j\omega\varepsilon_r\varepsilon_0\mathbf{E} \tag{1}$$

$$\nabla_T \times \mathbf{E} = -j\omega\mu_r\mu_0\mathbf{H} \tag{2}$$

where, ε_0 and μ_0 are the permittivity and permeability of free space, respectively. ε_r and μ_r are the relative permittivity and relative permeability, respectively. By substituting Eq. (1) into Eq. (2), it yield the vector Helmholts equations

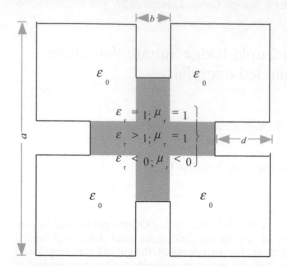

Figure 1. Dielectric loaded quadruple-ridge square waveguide.

$$\nabla_T \times \left(1/\varepsilon_r \, \nabla_T \times \mathbf{H}\right) - K_c^2 \mu_r \mathbf{H} = 0 \qquad (3)$$

where, ∇_T represents the transverse Laplacian operate and K_c^2 represents the cut-off wave number of the quadruple-ridge square waveguide. Since the Galerkin formulation is adopted, each set of weighting functions is equal to the corresponding set of basis functions. Assume $\mathbf{W}_i = \mathbf{W}_{ti} + W_{zi}e_z$, we have

$$h_t = \sum_{i=1}^{N_t} h_{ti} \mathbf{W}_{ti} \qquad (4)$$

$$h_z = \sum_{i=1}^{N_t} h_{zi} W_{zi} \qquad (5)$$

where, $\mathbf{W}_{ti}$ and $\mathbf{W}_{zi}$ represents the shape function of the element associated with interpolation edge i or node i. The summation index runs over the interpolation edges and nodes. In this paper, the QRSW is meshed into triangular elements. By applying Galerkin's method into Eq. (3), an eigenvalue equation can be obtained as

$$\begin{bmatrix} \mathbf{A}_{tt} & \mathbf{A}_{tz} \\ \mathbf{A}_{zt} & \mathbf{A}_{zz} \end{bmatrix}\begin{bmatrix} h_t \\ h_z \end{bmatrix} = K_c^2 \begin{bmatrix} \mathbf{B}_{tt} & 0 \\ 0 & \mathbf{B}_{zz} \end{bmatrix}\begin{bmatrix} h_t \\ h_z \end{bmatrix} \qquad (6)$$

where, $\mathbf{A}_{tt}$, $\mathbf{A}_{tz}$, $\mathbf{A}_{zt}$, $\mathbf{A}_{zz}$, $\mathbf{B}_{tt}$, and $\mathbf{B}_{zz}$ are known matrixes, while h_t and h_z are unknown vectors of the magnetic field. Eq. (6) can be written in simple matrix form

$$[\mathbf{A}][\phi] = K_c^2 [\mathbf{B}][\phi] \qquad (7)$$

where, $\mathbf{A}_{tt}$, $\mathbf{A}_{zz}$ and $\mathbf{B}$ are symmetric matrix. The full and necessary condition of having non-vanishing solution of Eq. (7) is it's coefficient determinant equal to zero, that is

$$\left|\mathbf{A} - K_c^2 \mathbf{B}\right| = 0 \qquad (8)$$

Eq. (8) is the characteristic equation and can be solved as generalized eigenvalue problems.

3 NUMERICAL RESULTS AND DISCUSSIONS

To verify the correctness of our code, the method is first used to calculate the cut-off frequency of dielectric loaded rectangular waveguide. The results are compared with the literature [3] and [8] and are given in Table 1. It is observed that the results of this paper are in good agreement with that of the references.

Next, the dispersion characteristics of QRSW with fixed aspect ratio $2d/a = 0.2$. Figure 2 shows the variations of the normalized cut-off wavelength λ_{c1}/a of the dominant mode with the ridge dimensions of air- or left-handed and right-handed materials. It is shown that in the case of right-handed materials loaded

a. λ_{c1}/a is considerably increased by increase in ε_r.
b. λ_{c1}/a increases continuously with $2d/a$ from 0.05 to 0.75.

In the case of left-handed materials loaded, it is shown that λ_{c1}/a is first increase then decrease with d/a. It is noted that λ_{c1}/a is much smaller than that of air or dielectric loaded case with the increase in $2d/a$.

The variations of the single-mode bandwidth $\lambda_{c1}/\lambda_{c2}$ with the ridge dimensions of different values of ε_r are shown in Figure 3. It is noted that $\lambda_{c1}/\lambda_{c2}$ is first increase then decrease with b/a. It is also

Table 1. Comparisons of the cut-off frequencies in the dielectric loaded rectangular waveguide.

	Cut-off frequency f_c/GHz				
Modes	This article	Ref. 8	Relative errors/%	Ref. 3	Relative errors/%
Dominant	1.8198	1.8151	0.25	1.7975	1.24
2nd-higher	3.4964	3.5445	1.35	3.5212	0.70
3rd-higher	3.6322	3.6347	0.06	3.6385	0.17
4th-higher	3.9802	3.9198	1.54	3.9972	0.42
5th-higher	4.0268	4.0632	0.89	4.0222	0.11
6th-higher	4.8715	4.8443	0.56	4.7789	1.93
7th-higher	4.9685	5.1969	4.39	4.9983	0.59

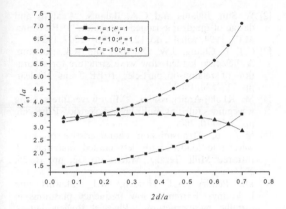

Figure 2. Normalized cut-off wavelength λ_{c1}/a of dominant mode versus $2d/a$.

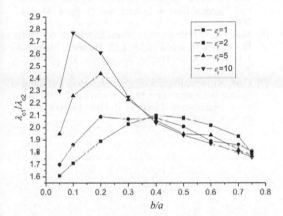

Figure 3. Single mode bandwidth $\lambda_{c1}/\lambda_{c2}$ versus b/a.

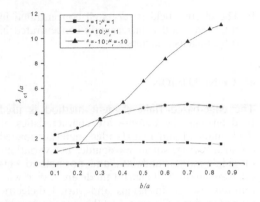

Figure 4. Normalized cut-off wavelength λ_{c1}/a of dominant mode versus b/a.

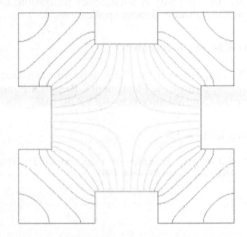

Figure 5. Dielectric loaded quadruple-ridge square waveguide $\varepsilon_r = 1$.

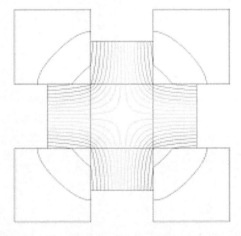

Figure 6. Dielectric loaded quadruple-ridge square waveguide $\varepsilon_r = 10$.

shown that $\lambda_{c1}/\lambda_{c2}$ increases rapidly as ε_r increases from 1.0 to 10.0.

Figure 4 shows the variations of the normalized cut-off wavelength λ_{c1}/a of the dominant mode for a fixed b/a for three cases: air loaded, dielectric with $\varepsilon_r = 10.0$ and $\mu_r = 1.0$ loaded, and LHM with $\varepsilon_r = -10.0$ and $\mu_r = -10.0$ loaded. It is shown that

a. λ_{c1}/a is considerably increased b/a for the case of LHM loaded.
b. λ_{c1}/a is first increase then decrease for the case of air—or right-handed materials loaded.
c. $\lambda_{c1}/\lambda_{c2}$ increases rapidly as ε_r increases from 1.0 to 10.0.

Figures 5 and 6 show the field patterns of the dominant mode in QRSW. It is shown that

a. The electric field is completely symmetrical along the line $a/2$.

71

b. The electric field is very well concentrated in ridge area and is increasingly concentrated in ridge area as ε_r increases.

4 CONCLUSION

The edge-based finite element method is used to determine the transmission characteristics of Right-handed and Left-Handed materials loaded quadruple-ridge square waveguide. The variations of single-mode bandwidth and normalized cut-off wavelength of dominant mode are shown with different ridge dimensions and varied dielectric loaded in ridge. Compared to the ridge waveguide loaded with air or conventional dielectric, the unusual behaviors of the LHM-loaded QRSW provide the possibility of novel device realizations for microwave and millimeter applications.

ACKNOWLEDGEMENT

The authors wish to acknowledge the assistance and support of the Fundamental Research Funds for the Central Universities—ZYZ2011056.

REFERENCES

[1] S.B. Cohn. "Properties of ridge waveguide," Proc IRE, no. 35, pp. 783–788, 1947.

[2] W. Sun, Balanis and C.A, Balanis, "Analysis and design of quadruple-ridged waveguide." IEEE Trans. MTT, 1994, vol. 42, no. 12, pp. 2201–2207, 1994.

[3] C.T.M. Chang, J.W. Dawson, and R.L. Kustom. A dielectric loaded slow wave structure for separation of relativistic particles, IEEE Trans Nuclear, pp. 526–530, 1969.

[4] M. Khalaj-Amirhosseini, "Microwave filters using waveguides filled by multi-layer dielectric," PIER, 66, pp. 105–110, 2006.

[5] Mai Lu, "Transmission characteristics of ridge waveguide loaded with left-handed materials," J. Infrared Milli. Terahz. Waves, no. 32, pp. 16–25, 2011.

[6] J.B. Pendry, A.J. Holden, W.J. "Stewart and I. Youngs, Extremely low frequency plasmons in metallic mesostructures," Physical Review letters, no. 76, pp. 4773–4776, 1996.

[7] D.R. Smith, W.J. Padilla, D.C. Vier, S.C., "Nemat-Nasser and S. Schultz, Composite medium with simultaneously negative permeability and permittivity." Physical Review Letters, no. 84, pp. 4184–4187, 2000.

[8] Mai Lu, "Transmission Characteristics of Ridge Waveguide Loaded with Left-Handed Materials" J Infrared Milli Terahz Waves, no. 32: pp. 16–25, 2011.

[9] C.M. Krowne, Physics of propagation in left-handed guided wave structures at microwave and millimeter-wave frequencies, Physical Review Letters, no. 92 053901, 2004.

Electrical, Control Engineering and Computer Science – Liu (Ed.)
© 2016 Taylor & Francis Group, London, ISBN 978-1-138-02937-8

A moving target collaborative tracking algorithm based on dynamic fuzzy clustering

Jing Xiong, Zhijing Liu, Guoliang Tang, Hongmin Xue & Jing Wang
School of Computer Science and Technology, Xidian University, Xi'an, China

ABSTRACT: In this paper, a collaborative tracking algorithm is proposed for settling the accuracy and energy consumption problems of moving targets tracking in wireless sensor networks that combined dynamic fuzzy clustering algorithm with behavior recognition solution. We can first subdivide the wireless sensor monitors into two types: the behavior recognition monitors and the collaborative tracking monitors, and then settle all the monitors by dynamic fuzzy clustering algorithm that utilizes the behavior recognition monitors as the original points. Each behavior recognition monitor is obliged to judge the behaviors of moving target and to wake up the corresponding collaborative tracking monitors using prediction and awakening mechanism aim at abnormal moving targets. Finally, we compare the solution with several existing methods in simulation environment, the statistics results show that the algorithm we proposed has a better tracking accuracy, which reduces energy consumption simultaneously.

Keywords: collaborative tracking; dynamic fuzzy clustering; Wireless Sensor Networks (WSN); behavior recognition

1 INTRODUCTION

Wireless sensor networks have been applied to widespread areas with the development of microelectronics embedded system and wireless communication technology. Wireless monitors cost low and can be settled easily. However, their communicating and computing abilities are limited. Since moving target tracking is an important application of wireless sensor networks, to improve the detection accuracy and tracking efficiency and to reduce the energy consumption become the aim targets and the measurements of collaborative tracking performance [3].

Researchers have proposed different solutions for various application scenes aim at collaborative tracking issue. The solutions hammer at raising the tracking accuracy and reducing energy cost. Wang Xinbo [1] proposed a collaborative tracking solution that combined maximum likelihood estimation with kalman filtering. Diluka Moratuwage [2] following static and dynamic points and put forward a method of collaborative location and tracking based on the multi movement tool that using random finite set to express map features. Zhu Guibo [4] combined short time tracker with long time target detector and proposed a dynamic collaborative tracking solution, which is good for frame losing, mutation and long time interference problems. Liu Qin [5] utilizing binary particle swarm optimization algorithm for tracking multi hiding targets in radar networks.

Chen Chaochun [6] presents a three layers architecture Hama to achieve targets collaborative tracking and location management. X. Xing [7] depending on three status transport module, put forward a moving target tracking method based on drove, experiment result which shows that the solution play a good performance in detection, awakening mechanism and logical networks architecture.

In this paper, we divide monitors into groups by function and utilize dynamic fuzzy clustering algorithm for collaborative tracking, the behavior recognition monitor is obliged to judge the behaviors of moving target and to wake up the corresponding collaborative tracking monitor using prediction and awakening mechanism aim at abnormal moving targets. The experiment results show that the algorithm we proposed has better detection and tracking accuracy, less energy consumption than the solutions we mentioned above.

2 RELATED WORKS

2.1 *The composition of WSN*

WSN (Wireless Sensor Networks) is wireless network that some collection messages are added to, such as videos, audios, images. In this paper, we refer to two kinds of monitor nodes: Behavior Recognition Monitor (BRM) which is expressed as circles in Figure 1 is used to recognize behaviors

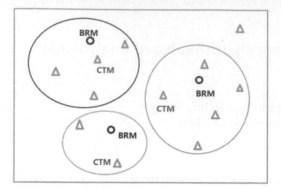

Figure 1. The Sketch map of composition of WSN.

of moving targets and manage messages, Collaborative Tracking Monitor (CTM) that is expressed as triangles in Figure 1 is responsible to collaborative tracking aim at the targets that BRM locked. The hardware of each monitor node is composed of three parts: video collection module, intelligent message processing module and wireless communication module.

The guard areas of BRM and CTM are multiplicity for guarantee the targets recognition and tracking accuracy rate. The difference of work mode between the two kinds of monitor nodes is that the BRM is always active but the CTM has two work modes: sleeping mode and active mode. So the WSN could track multi-targets and make the rest node to be sleeping mode to save energy consumption.

2.2 Dynamic fuzzy clustering algorithm

In our system, a dynamic clustering algorithm based on FCM algorithm is proposed which original clustering centre is confirmed, and we could get the groups consist of BRM and CTM, each monitor node complete the behavior recognition and relay tracking task in the same group.

Although FCM algorithm is good for fast clustering and could get the better clustering accuracy, it fall into terminal point easily, so the choosing for original point is important. Considering the features of our system, let BRM as the original clustering centre. We assume $X = \{x_1, x_2, ..., x_n\}$ as the total monitor nodes set, divide the X into c groups ($2 \leq c \leq n$), there are c number of BRM as original clustering centre $V = \{v_1, v_2, ..., v_c\}$ at the same time. The degree of membership for each monitor node could be described as fuzzy matrix $U = [u_{ij}]$, u_{ij} is the degree of membership between the monitor i and j, that must satisfy the conditions $\sum_{i=1}^{c} u_{ij} = 1$ and $u_{ij} \in [0,1]$. The destination function could be described as [10]:

$$J(U, v_1, ..., v_c) = \sum_{i=1}^{c} J_i = \sum_{i=1}^{c} \sum_{j}^{n} u_{ij}^m d_{ij}^2 \qquad (1)$$

In the equation above, v_i is the clustering centre in the fuzzy group i, $d_{ij} = \|v_i - x_j\|$ is the Euclidian Distance between the i clustering centre and the j monitor node. M $m \in [1, \infty)$ is the fuzzy weighting index that benefits for noise restraining and making membership function smoothing. Two conditions contribute to get the lowest value for destination function mentioned above:

$$v_i = \frac{\sum_{j=1}^{n} u_{ij}^m x_j}{\sum_{j=1}^{n} u_{ij}^m} \qquad (2)$$

$$u_{ij} = \frac{1}{\sum_{k=1}^{c} \left(\frac{d_{ij}}{d_{kj}}\right)^{2/(m-1)}} \qquad (3)$$

The degree of membership m controls the share degree of fuzzy groups, it also play an important role in noise restrain and effect the concave-convex of destination function. Considering the convergence of the algorithm, the value area of m is related with sampling amount X, the best value is in the range [1.3, 2.5]. Though the algorithm that set the confirmed original clustering point, we can get the cluster result which BRM are the cluster centre points. In each group according to the result includes a BRM and several CTM, CTM receives the tracking command from BRM in the same group, and product a dynamic monitor subnet. When the CTM in the boundary of a group is busy, the WSN will dynamic cluster for expanding the radius of the busy subnet according to a clustering intervene factor β to share responsibility for busy subnet.

2.3 Recognition of anomalous targets

In order to save energy consumption and improve tracking efficiency, BRM and CTM divide the task. BRM is responsible for recognizing the behaviors of moving targets in some important place. We can

Figure 2. The jumping behavior template.

74

set some legitimate behavior according to the need in fact. For example, we define walking and slow running as the normal behaviors, so other actions are anomalous behaviors. CTM in the same group only need to track the suspicious target related.

Our solution recognize behavior of motion target utilize the method couple template matching with feature points tracking [9]. Each monitor in the WSN share the same behavior library which some behavior template would be defined, such as walking, running, jumping. Surveillance system can definite some regular behaviors in the beginning, so other behaviors are suspicious. For example, the jumping behavior could be shown as the behavior template (Fig. 2).

The similarity of the two images could be defined according to PMSD (Procrustes Mean Shape distance). The type of behavior would be conjecture through minimum standard variance algorithm aim at behavior sequence. The standard variance is used to describe the data that behalf the degree of deviation from average data, standard variance increase with the degree of deviation and vice versa. The similarity distance matrix between standard behavior sequences with the real-time video could be present with [8]:

$$S_j^k = \min_{i=1,2\ldots8} (d_{(j)}^{(i)}(\hat{u}_1,\hat{u}_2)) \tag{4}$$

So BRM could judge whether or not the current motion targets should be collaborative tracking. In this solution CTM only need to collaborative the suspicious motion targets, so it decrease the burden of system enormously and avoid a large calculation for paralleling tracking and behavior recognition.

2.4 Prediction and activation mechanism

To further save energy consumption and take full advantage of the result utilizing dynamic fuzzy clustering algorithm, we propose the prediction and activation mechanism in the system. BRM is the leader of dynamic subnet, so it always active, but CTM is on the default sleeping mode. CTM would be activated when the target need to be collaborative tracking according to the trajectory.

The method to predict trajectory of moving target and to activate corresponding CTM is a dynamic process as the monitor subnets are dynamic clustered. BRM take charge of transmission the order ACTIVE to CTM expect recognize the targets behaviors, so it must maintain a latest subnet member list and boundary messages for the subnet in the all directions. When a suspicious target move in a monitor subnet, the BRM would predict the trajectory of the target and send order ACTIVE to CTMs related. If the target moves to

the boundary of the subnet, BRM will send COWORK order to the BRMs nearby that would active CTMs in the same subnet to relay tracking the target. The COWORK message includes the current subnet number and features of the moving target such as moving speed. When the tracking task completed or target is not through the scope of CTM for a long time, it will change to sleep mode automatically for saving energy.

3 EXPERIMENT AND ANALYSIS

For demonstrating the veracity and efficiency of the solution, we analysis the result of simulation experiment and compare it with several similar algorithms existed. We assume the supervisory control area is 20*30 m that shown in Figure 3, 5 BRM for behavior recognition and 22 CTM for collaborative tracking random distribution in it. A moving target uniform move from point A to B on the speed $v = 1.2$ m/s, the degree of membership value using dynamic fuzzy cluster is $m = 2$, the energy consumption for receiving circuit and emission circuit in each monitor could be valued as $e_s = e_r = 60$ nJ/bit. Suppose the trajectory of moving target is just like the curve in Figure 3.

The result of dynamic clustering can be shown in Figure 3 using the solution we proposed. When the

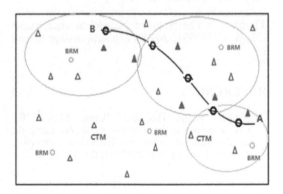

Figure 3. Moving target collaborative tracking by WSN.

Table 1. The result of statistics analysis for different algorithms.

Solutions	Detection accuracy	Tracking accuracy	Energy wastage/J
Xinbo Wang [1]	84.6%	79.1%	0.631 e8
Diluka Moratuwage [2]	78.5%	81.4%	2.135 e8
Our solution	85.7%	86.9%	0.574 e8

moving target accesses the guard area from point A, it will be detected by the BRM in lower right corner and ascertained as collaborative target. The CTMs described with solid triangles would be activated by the BRM according to the location that the target be and relay tracking the target until it leaves the whole guard area from point B.

Table 1 has shown the comparison of different methods with our method aiming at detection accuracy, tracking accuracy and energy wastage. The results show that our solution has a better performance than the similar solutions that we mentioned.

4 CONCLUSION

In this paper, we proposed a collaborative tracking algorithm to work out the accuracy and energy consumption problems of moving targets tracking in wireless sensor networks that combined dynamic fuzzy clustering method with behavior recognition algorithm. Compared with the solution with several existing methods, the results have shown that the algorithm we proposed have better detection and tracking accuracy, reducing energy consumption simultaneously.

ACKNOWLEDGEMENTS

Thanks for The National Natural Science Funds (Num: 61173091, 61202177) supporting to our research.

REFERENCES

[1] Xingbo Wang, Huanshui Zhang, Minyue Fu. Collaborative target tracking in WSNs using the combination of maximum likelihood estimation. J Control Theory Appl 2013 11(1)27–34.

[2] Diluka Moratuwage, Ba-Ngu Vo and Danwei Wang. Collaborative Multi-Vehicle slam with Moving Object Tracking. 2013 IEEE International Conference on Robotics and Automation, 5702–5708.

[3] Oualid Demigha, Walid-Khaled Hidouci, Toufik Ahmed. On Energy Efficiency in Collaborative Target Tracking in Wireless Sensor Network: A Review. IEEE Communications Surveys & Tutorials, Third Quarter 2013, Vol. 15, No. 3, 1210–1222.

[4] Guibo Zhu, Jinqiao Wang, Changsheng Li, Hanqing Lu. Collaborative Tracking: Dynamically Fusing Short-Term Trackers and Long-Term Detector. 2013, Part two, LNCS 7733, pp. 457–467.

[5] Qin Liu, Zheng Liu, YunFo Liu, Rong Xie. Maneuvering target collaborative tracking algorithm with multi-sensor deployment optimization System Engineering and Electronics. 2013.Vol. 35 No. 2.

[6] C.-C Chen, J.-M. Hsu, and C.-H. Liao, Hama: A Three-Layered Architecture for integrating Object Tracking and Location Management in Wireless Sensor Networks. Third International Conference on Multimedia and Ubiquitous Engineering, 2009, pp. 268–275.

[7] X.Xing, G. Wang, and J. Wu, Herd-based Target Tracking Protocol in Wireless Sensor Networks, Wireless Algorithms, Systems, and Applications, 2009, pp.135–148.

[8] Zhang Jun, Liu Zhijing, Research on Analysis and Recognition about Abnormal Behavior of Moving Human in Video Sequences, Doctoral 2009, chapter 3.

[9] Zhang hao, Liu Zhijing and Zhao Haiyong, Human Activities for Classification via Feature Points. Information Technology Journal 10(5);974–982. 2011.

[10] Jiang Lun, Ding Hua-fu, Improvement of the Fuzzy C-Means Clustering Algorithm, Computer & Digital Engineering. Vol. 38 No. 24, p 4–6,14.

Electrical, Control Engineering and Computer Science – Liu (Ed.)
© 2016 Taylor & Francis Group, London, ISBN 978-1-138-02937-8

Research on the bacon key-quality based on physical detection and hyperspectral image of full-scale features

Pei-yuan Guo, Man Bao, Kun-cheng Yang & Shuo Liu
School of Computer Information Engineering, Beijing Technology and Business University, Beijing, China

ABSTRACT: Detection techniques of computer vision and NIR as well as hyperspectral image were investigated in this paper. Hyperspectral image detection technology for bacon quality detection has become a research hotspot recently. The basic principle of using hyperspectral image detection technology and the research status in bacon quality detection at home and abroad are introduced in the paper. The prospect of future research in bacon quality detection using hyperspectral image detection technology is proposed to researchers on the related study as a reference.

Keywords: bacon quality; nondestructive detection technique; hyperspectral image detection

1 INTRODUCTION

Bacon is a typical representative of Chinese traditional meat products with its long history and unique flavor. It is an important part of the world precious food culture heritage. Because of its special composition and structure, bacon is easily microbial spoiled, such as volatile basic nitrogen TVB-N (metamorphic), the degree of oxidation of TBA, the total number of colonies, acid value and peroxide value, the nitrite residue, protein, fat, and other quality, all of which exceed the standard. The traditional sensory detection was influenced by subjective factors. Chemical test requires a series of chemical means to complete the real-time detection of complex process, which was long detection time and not rapid[1]. Detection limit of time is much too long to cause the bacon sausage spoilage or depreciation. Therefore, considering China's food safety inspection technique status, there is an urgent need for a rapid and accurate practical detection technique for bacon quality[1], so as to determine whether it can be eaten. Hyperspectral image testing technology used in nondestructive testing of bacon quality was drawn attention.

2 BACON QUALITY NONDESTRUCTIVE TESTING TECHNOLOGY

2.1 *Artificial olfactory and artificial taste detection technology*

Various microbial contamination or corruption was caused by the decomposition of the enzyme and its sour nature of the fermentation when Bacon was deteriorated. It will produce sulfide and ammonia.

Therefore, we can use Artificial Olfactory System (AOS)[2] to test the variation of the time of sulfide and ammonia concentrations, then to determine the freshness of Bacon. Currently, the study that uses artificial olfactory system to test the freshness of Bacon is quite extensive abroad. Domestic experts began to apply artificial olfactory system to detect the freshness of Bacon. For example, Guo Peiyuan and Qu Shihai[3] made a set of intelligent detection recognition system based on electronic information technology, optical detection technology, image processing, and neural network pattern recognition technology. It collected ammonia and hydrogen sulfide released from the process of pork corruption, and collected data as a set of neural network input. It is a reliable basis for testing the freshness of pork[4]. Olfaction visualization technology is currently a new branch of artificial olfactory research. It is able to solve some common problems in gas sensors, and compensate for certain deficiencies in the electronic nose system. Its outstanding feature is translating smell information into visual information, so that smell is "visible." Olfaction visualization technology was first proposed by Professor Kenneth S. Suslick from the University of Illinois at Urbana—Champaign. It is a visualization method for qualitative and quantitative analysis based on color change of gas to be detected after reaction with chemical reagent[5].

Artificial Taste System (ATS) technology started not long ago, the technology is yet to be further developed. The combination of artificial olfactory and artificial taste technology will further improve the detection and recognition of food, but the related research is in its infancy.

2.2 Computer vision technology

Computer vision testing technology is a practical technology. It can get the object image with the image sensor (commonly used high-resolution CCD), then converts the image into digital image. Image processing and image analysis are the core of computer vision technology[6-7].

Computer vision system has the advantages of nondestructive, reliable and fast, so it is widely used in meat quality testing, not only in the visible area, but extends to the near infrared, and other areas, Zhang Zhe[8] once detected the fat content of pig eye muscle intramuscular with computer vision technology which is also used to study fluctuations and grading of kidney meat quality. For example, we can detect a clear PSE kidney meat with the 400~700 nm optical reflection measurement system.

2.3 Near infrared spectroscopy technology

Near infrared spectroscopy technology has been a new optical detection technology, which has been widely used recently in the food industry. The technology was integrated with spectroscopy, Chemical measurement, computers and other multidisciplinary modern analytical techniques. It has the advantage of rapid food online analysis, Pollution-free, nondestructive analysis and testing, which can perform remote analysis, detection and so on. It is not only able to detect the chemical composition of meat in the traditional ways, but also can include sensory quality evaluation, species identification and other aspects of meat quality. It has become a very active research area. Because of the majority of organic compounds containing different hydrogen groups in meat, the content of these components can be determined by near infrared spectroscopy. And through further analysis, we can get more information related with the meat quality[2]. There are many domestic and foreign experts using near infrared technology for meat detection. For example, it has been successfully achieved detecting the water changes of meat in the heating process with near infrared fiber optic probe. It provides a new way of effective and reasonable control processing for meat industry. However, near infrared spectroscopy also has its fatal flaws, such as requiring a large representative sample of chemical values to make model, low accuracy of quantitative analysis, complicated calculation model established, and so on.

3 HYPERSPECTRAL IMAGE DETECTION TECHNOLOGY OF BACON QUALITY

Most of the current nondestructive testing methods use only a single method of detection, or traditional signal preprocessing techniques and pattern recognition methods. The method often has information response of only one or two indexes, but does not evaluate comprehensively varieties of information. Since overall quality of bacon is complex, it should be evaluated comprehensively by a number of indicators. Therefore, for the organic integration of these detection techniques, taking full advantage of multiple information, simply, quickly, accurately, and comprehensively evaluating bacon is the focus of future research and development.

Hyperspectral image detection technology is more and more popular research at home and abroad. Spectroscopy technique can detect bacon material structure, composition, and other internal quality information. Computer vision technology can fully reflect the external characteristics of meat. Therefore, hyperspectral images can reflect the overall quality of bacon.

3.1 Hyperspectral image detection technology

As a new detection technology, hyperspectral, image technology focus optics, optoelectronics, electronics, information processing, computer science, and other fields of advanced technology are combined with the traditional two-dimensional imaging techniques and spectroscopy. The technology has a characteristic of super multiband, high spectral resolution, and one map. So hyperspectral technology has a greater detection advantage and detection accuracy in agricultural and livestock products, food quality, and safety testing. Hyperspectral imaging technology has been used in fruit internal quality, surface contamination and bruises testing, vegetable maturity and internal quality testing, and will be further applied to the internal quality of the bacon inspection.

The hardware components of hyperspectral image detection include light source, CCD camera, and a computer equipped with image acquisition card as shown in Figure 1. Three-dimensional hyperspectral image data of bacon can be obtained, as shown in Figure 2.

At home and abroad, although there have been some related research reports in hyperspectral image detection of meat quality, there are no related research reports based on hyperspectral image detection of **bacon** quality of the test.

Guo Peiyuan, Fu Yan and other researchers are studying the relevant issues about bacon quality test with hyperspectral image detection technology. Hyperspectral images have been obtained in the 700–1500 nm wavelength range of 512 bands. We can get the effective characteristic wavelength with principal component analysis. For example, bacon

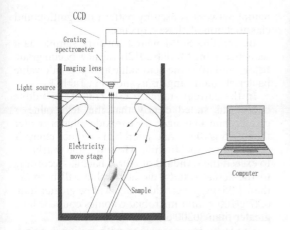

Figure 1. Hyperspectral imaging system.

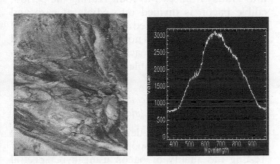

Figure 2. 3d hyperspectral imaging data of bacon.

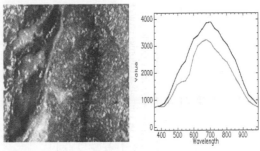

Figure 3. The plaque of hyperspectral component analysis.

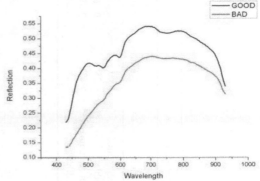

Figure 4. The position of plaque and wavelength curve.

hyperspectral images were collected by the hyperspectral imaging system, which was carried out by principal component analysis. Images can be clearly some parasites in the sixth principal components band image, which is shown in Figure 3. From the figure, we can clearly identify the plaque location of hyperspectral images after principal component analysis. Figure 4 shows the location of no plaque and plaque wavelength curve fitting graph. Red line in this figure is the position of the waveform to nonbacteria, and blue line is the position of the waveform of bacteria. As a result, the study to detect bacon plaque with hyperspectral image detection is feasible and has practical significance.

At the same time, the researchers also studied the detection of bacon quality and related issues with hyperspectral image detection technology. Hyperspectral images are got in the near infrared range. Bacon samples in near infrared range images within a band are shown in Figure 4. It can be seen from the figure that large amount of reflectivity information can be contributed to the follow-up analysis.

3.2 Bacon food grade division

The acid value and peroxide value of bacon are the main standard testing quality in our country and reflect the difference between edible and inedible. In the current standard GB 2730-2005, only acid value and peroxide value of bacon will be divided into 2 types and will not exceed the standard. But the experiment shows that although acid value and peroxide value do not exceed the standard of GB of bacon, sensory characteristics are obvious, such as sour, sticky. Then, detecting bacon obviously cannot meet the growing needs of people for food quality requirements.

At the same time or in the storage environment influence conditions, although the acid value and peroxide value of bacon cannot meet the standard, microbial colonies number has been exceeded.

In bacon spoilage process, the acid value and peroxide value and the high spectrum can be obtained. In order to improve the classification accuracy of bacon food grade division, total information of microscopy image and spectral image are needed to carry out multiple data fusion. SOM

Table 1. The bacon edible grade redrawing.

Before grade			After grade		
Grade	Number	Acid value (mg/g), Peroxide value (g/100 g) and the total number of microbial colonies (cfu/g)	Grade	Number	Acid value (mg/g) and Peroxide value (g/100 g) and the total number of microbial colonies (cfu/g)
Edible	118	$<=3.84; <=0.23; <=29500$	Rest assured	40	$<=0.14; <=0.13; <=9500$
Inedible	32	$>=4.05; >=0.27; >=30200$	Edible	35	$0.16–2.28; 0.15–0.18; 10600–22000$
			Not recommended	43	$2.65–3.84; 0.19–0.23; 23800–30000$
			Inedible	32	$>=4.05; >=0.27; >=30200$

neural network is used by pattern recognition and classification of these sequences.

By training SOM neural network sample data sequence in the Matlab 2012b, the new bacon grade before and after the acid value and peroxide value bacterial count range is as shown in Table 1.

In the current standard GB 2730-2005, Edible bacon has stated clearly that the acid value is greater than 4.00 mg/g, the peroxide value is greater than 0.25 g/100 g, and microbial colonies count is greater than 30000 cfu/g. However, the market is to exceed the standard of edible bacon. According to the actual results, the acid value will be more than 4.05 mg/g, peroxide value will be greater than 0.27 g/100 g, and microbial colonies count will be greater than 30200 cfu/g.

Similarly, in accordance with national standards GB 2730-2005, bacon food can be divided into grades from two to four. During the period, edible bacon is divided into edible at ease and edible. Inedible is divided into not recommended and inedible. The results not only meet the national standard need, but also achieve fine division purpose of edible and inedible bacon.

3.3 The prospect of bacon quality hyperspectral image detection technology

Hyperspectral image detection combines the advantages of spectroscopy and imaging study. Testing the quality of bacon using hyperspectral image detection technology is improved in recent years. Through a bacon detection of preliminary study, this paper shows that hyperspectral imaging technology of Inside and outside **Bacon** quality can be analyzed and researched. Hyperspectral imagery for bacon quality testing has had good results, research ability and feasibility.

Hyperspectral imaging system has obvious advantages in determining the characteristic wavelength of the bacon quality parameters. Through researching and testing, we can determine the most effective characteristics wavelength of bacon quality. In this way, we can design a number of wavelengths of spectral imaging systems and then apply it to the actual production. It can greatly improve the detection efficiency, achieve the purpose of online detection, rapid and non-destructive testing.

4 CONCLUSION

With the development of science and technology, hyperspectral image detection technology can obtain extensive images and spectral information of testing meat. It can nondestructive test meat quality and comprehensive evaluation, so it has great potential for development in all kinds of meat quality testing.

Light absorption characteristics of different meat are not the same. The advantage of hyperspectral image detection is that it can choose the best effective characteristic wavelength according to the test target. So this method increases the potential applications of hyperspectral image detection technology. To extend the field of apply hyperspectral image detection technology, other important meat qualities can be studied, such as freshness, drug residues, heavy metals, processing aids, additives, and so on. Nondestructive testing technology of meat quality and especially hyperspectral image detection technology in China are still in the experimental research stage, and do not really put them into actual production yet. Therefore, it has a very important practical significance to the modern development of meat industry, and it accelerates the market application of meat quality nondestructive testing equipment, and achieves meat online, fast, nondestructive testing.

This undergoing project was supported by the Natural Science Foundation of China (No. 61473009), the Natural Science Foundation of Beijing (No. 4122020), and the Beijing Board of education and scientific research (PXM2015_014213_000063) *Corresponding author: Guo Pei-yuan.

REFERENCES

[1] Monin G., "Recent methods for predicting quality of whole meat," J. Meat Science. vol. 49(2), 1998, p. 231.

[2] Xu Xia, Cheng Fang, and Ying Yi-bin, "Application and Recent Development of Research on Near-Infrared Spectroscopy for Meat Quality Evaluation," J. Spectroscopy and Spectral Analysis. vol. 29(7), 2009, pp. 1876–1880.

[3] Xu Shi-hai and Guo Pei-yuan, "Study on detection method to fresh degree of meat based on many message dealing," J. Journal of Beijing Technology and Business University (Natural Science Edition). vol. 24(5), 2006, pp. 26–31.

[4] Xu Guan-nan, Guo Pei-yuan, and Yuan Fang, "Development of Nondestructive Detection Techniques of Pork Freshness," J. Journal of Beijing Technology and Business University (Natural Science Edition). vol. 28(1), 2010, pp. 14–17.

[5] Zou Xiao-bo and Zhao Jie-wen, "Nondestructive testing technology of agricultural products and data analysis method," M. Beijing: Light Industry Press of China. pp. 1–8, 2008.

[6] Chen Chun, "Computer image processing technology and algorithm," M. Beijing: Tsinghua University Press. pp. 3–15, 2003.

[7] Li Ming-jing, Liu Guan-yong, Zhang Zhi-wei, "The computer vision based beef automatic grading technology," J. Meat Science. vol. 6, 2007, pp. 18–20.

[8] Zhang Zhe, "The computer vision technology based the fat content determination of pig eye flesh," J. Swine Industry Science. vol. 2, 2006, pp. 24–25.

Electrical, Control Engineering and Computer Science – Liu (Ed.)
© 2016 Taylor & Francis Group, London, ISBN 978-1-138-02937-8

Study of eddy current brake based on motion of Permanent Magnet in the nonmagnetic metal tube

Yuxuan Xia, Songming Yan, Haining Tan & Wei Song
Tianjin No. 1 High School, Tianjin, China

Siyu Yang & Wei Chen
School of Electrical Engineering and Automation, Tianjin University, Tianjin, China

ABSTRACT: Eddy current, which hinders the movement of the PM, will be induced in a non-magnetic metal tube when a Permanent Magnet (PM) moves in the non-magnetic metal tube. In this paper, equivalent current model and the law of electromagnetic induction are adopted to analyze the magnetic field around the PM and the eddy current in the metal tube respectively. With the help of Finite Element Method, permanent magnetic field and eddy current distribution are calculated. Besides, a discrete calculation method is proposed to complete numeric calculation of the electromagnetic force. Furthermore, experimental platform is established and experiments, which demonstrate the relation between velocity of PM and electromagnetic force and the influence of metal resistivity on the electromagnetic force, are designed.

Keywords: Permanent Magnet; non-magnetic metal; eddy current effect; electromagnetic braking

1 INTRODUCTION

According to the electromagnetic theory, the induced electromotive force and eddy current are produced in the metal tube to prevent the movement of Permanent Magnet (PM) when PM moves in the metal tube. Eddy current results in electromagnetic power loss and heating of the conduct. Meanwhile, eddy current also produces electromagnetic force applied to the PM.

Eddy current effect is often applied in the electromagnetic mechanism. By the heat effect of eddy current, high frequency current is injected into the smelting device in the metal smelting process, which produces large eddy current in the smelting metals and the heat generated by the vortex can make the metal melt quickly. Meanwhile, eddy current effect can also cause some negative effects. Eddy heat needs to consume additional energy, which reduces the efficiency of the electromagnetic mechanism, and eddy heat may shorten the service life of insulating material, and declines the electromagnetic properties of materials.

This paper studies the physical phenomenon that PM subjects to electromagnetic force braking when it moves in nonmagnetic metal tube. The conclusion can provide certain reference for taking advantage of eddy current magnetic field in the process of buffer braking.

2 ANALYSIS AND MODELING

Now make the following assumptions: (1) The PM is uniform magnetized along the axial direction. (2) Place the metal tube vertically, and the direction of magnetic pole of PM coincides with the symmetry axis of the tube. (3) The PM falls down vertically along the axial line of the tube without the translation and deflection in the horizontal direction. (4) The metal tube is nonmagnetic, and the distribution of PM magnetic field in the tube is the same as that of vacuum environment. (5) The relative permeability of PM equals 1.

2.1 Analysis of PM magnetic field

A right-handed three-dimensional coordinate system is built in Figure 1, regarding the center of cylindrical PM as origin and the PM N-pole direction as the z axis positive direction. In the equivalent current model of the PM, the magnetic field is expressed as distributed current element with surface current density of $J_{sm} = M \times n$. The magnetic vector potential A can be expressed as

$$A = \frac{\mu_0}{4\pi} \oint_S \frac{J_{sm}}{r} dS \qquad (1)$$

where S is the surface surrounding PM, r is the distance from source point to field point, μ_0 is the

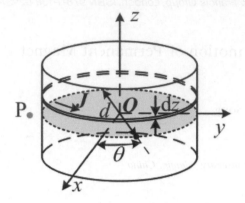

Figure 1. Three-dimensional coordinate of PM.

air permeability, M is the magnetization intensity of PM, n is the unit normal vector on the outer surface of PM.

According to the field superposition principle, the magnetic induction intensity at P produced by the whole cylindrical surface current is

$$B = \nabla \times A = \int_{-h/2}^{h/2} dB_l = \frac{\mu_0}{4\pi} \int_{-h/2}^{h/2} \oint_l \frac{Mdl \times r}{r^3} dz \quad (2)$$

We consider two current elements ($r\cos\theta$, $r\sin\theta$, 0) and ($r\cos\theta$, $-r\sin\theta$, 0), with corresponding element current vectors of respectively ($-M\sin\theta$, $M\cos\theta$, 0) and ($M\sin\theta$, $M\cos\theta$, 0). The synthesis of magnetic induction intensity produced by two symmetrical current elements at P (x_1, 0, z_1) is

$$dB = \frac{\mu_0}{2\pi} \frac{(z_1 M\cos\theta, 0, 0.5dM - x_1 M\cos\theta)}{(x_1 - 0.5d\cos\theta)^2 + (0.5d\sin\theta)^2 + z_1^2} dl \quad (3)$$

It's not difficult to find out from Eq. 3 that the resultant magnetic field at point P produced by any two current elements symmetrical relative to the plane xOz don't have component on the y axis. In conclusion, according to the field superposition principle and the symmetry of the ring line current, the magnetic induction intensity generated by the cylindrical surface current of the PM also doesn't have tangential component. Only the radial component B_r and vertical component B_z of magnetic induction intensity exist.

2.2 Analysis of eddy current in metal tube

When the PM moves with velocity v, the motional electromotive force induced in the closed loop of metal tube is expressed as

$$\varepsilon = \oint_l (v \times B) \cdot dl = \oint_l (-vB_r\sin\theta, vB_r\cos\theta, 0) \cdot dl \quad (4)$$

It's not difficult to find out from Eq.4 that ε is only affected by B_r. Therefore, the vertical component B_z is not considered during the analysis of eddy current field. With the symmetry of the magnetic field and metal tube, the analysis of electromagnetic induction is carried out in the vertical cross section through center axis of the metal tube as shown in Figure 2. For a random point P (x, z), its area is expressed as dS, its thickness is expressed as dl, and its resistance and resistivity are expressed respectively as dR and ρ. So the eddy current density is

$$J(x,z) = \frac{1}{dS} \frac{d\varepsilon}{dR} = \frac{B_r(x,z)v}{\rho}. \quad (5)$$

In conclusion, in any vertical cross section through the center axis of the metal tube, the direction of the eddy current density produced by electromagnetic induction is vertical to the cross section. Extending the two-dimensional analysis results to three-dimensional area, it can be obtained that the induced current density J in the tube only has tangential component J_θ in any horizontal plane, and there is no vertical component J_z. Hence, the direction of eddy current is along with the tangent direction of the metal tube.

2.3 Modeling of electromagnetic force

As shown in Figure 2, the force of the PM generated by the magnetic field of eddy current is

$$F_e = -\iint_{Sa} \oint_l dl \times B_r(x,z)J(x,z)dS = k\frac{2\pi v}{\rho} Z \quad (6)$$

where $Z = \iint_{Sa} rB_r^2(x,z)dS$, Sa is half the area of cross section, k is the unit vector of z axis.

It can be seen from Eq.6 that the direction of force is directed along the positive z axis, which is reverse direction relative to the velocity v, so it is a braking force, its value is related to the B_r, the relative velocity v, resistivity ρ and the area Sa.

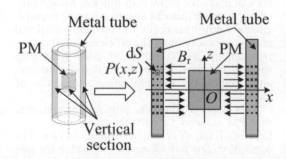

Figure 2. Vertical cross section of metal tube.

3 ELECTROMAGNETIC CALCULATIONS

3.1 Calculation of magnetic field

The PM material is set to be Nd-Fe-B, its shape is set to be cylinder. The parameters needed in Finite Element calculation are listed as shown in Table 1.

The cylinder center is the origin and the central axis of the PM is z axis. The three-dimensional model meshed by tetrahedron grid and the magnetic field distribution of PM are shown in Figure 3. It can be seen that the static magnetic field distribution of PM is symmetrical and there is only radial and vertical component but no tangential component.

In the above established model of PM, 10 adjacent vertical straight line with even interval of 1 mm with distance of 12 mm ~ 16 mm from the axis of PM are chosen in XOZ plane, and the static magnetic field of every line are calculated respectively as shown in Figure 4. Five lines located on the right wall of tube are marked with l_1, l_2, l_3 l_4, and l_5. The tangential component B_θ, radial component B_r and vertical component B_z on $l_1 \sim l_5$ are shown in Figure 5.

It can be seen that the tangential component B_θ is always zero. The radial component B_r is an odd function. Its direction changes in the midpoint of each line and its amplitude reaches maximum on the positions of upper and lower surfaces of the cylindrical PM. The vertical component B_z is an

Table 1. Parameters for FE calculation.

Items	Value
Magnet height h	30 [mm]
Magnet diameter d	20 [mm]
Remanent magnetic density B_m	1.4 [T]
Inner diameter of metal tube D_{in}	26 [mm]
Outer diameter of metal tube D_{out}	32 [mm]
Length of metal tube s_m	500 [mm]

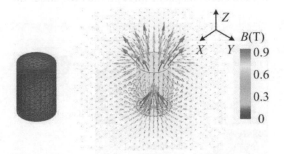

Figure 3. Mesh generation and magnetic field distribution.

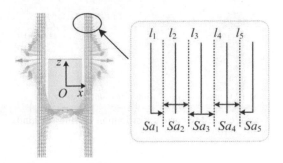

Figure 4. Magnetic field distribution on vertical section.

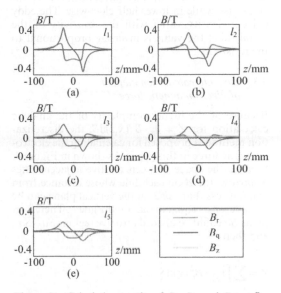

Figure 5. Calculation results of B_θ, B_r and B_z on five lines.

even function and its direction changes on the upper and lower surfaces of the cylindrical PM.

3.2 Calculation of eddy current distribution

Make the PM falls through T2 copper tube, H62 brass tube and 6061 aluminum alloy tube with the same sizes in Table 1. The resistivity of three kinds of metal tubes are $\rho_{T2} = 0.018 \times 10^{-6}$ Ω/m, $\rho_{H62} = 0.071 \times 10^{-6}$ Ω/m, $\rho_{6061} = 0.04 \times 10^{-6}$ Ω/m, respectively. The eddy current produced in the metal tube is calculated by transient field solver. As shown in Figure 6, eight vertical lines evenly distributed along a cylindrical surface which have a distance of 14.5 mm from the center axis of the tube are selected to analyze the eddy current density.

It can be seen from Figure 6, the eddy current distribution is symmetrical relative to the

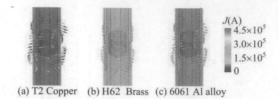

(a) T2 Copper (b) H62 Brass (c) 6061 Al alloy

Figure 6. Eddy current distribution in different kinds of metal tubes.

horizontal center plane of the metal tube, and the eddy current is directed along the tangent line. In upper half of the PM, eddy current density is anti-clockwise while in lower half clockwise. The eddy current density varies with the resistivity of the metal tube; the value is inversely proportional to the resistivity.

3.3 Discrete calculation method of electromagnetic force

Because of the higher complexity of the analytic calculation for $Z = \iint_{Sa} r B_r^2(x,z)\mathrm{d}S$, the discretization method is proposed for calculating the electromagnetic force in this paper. As shown in Figure 4, because the space between the five adjacent lines is narrow, the B on each line whose distance from line l_1 is less than $\Delta h/2$ on the vertical plane can be considered as equal to that on the line l_1. Therefore, plane Sa could be divided into five regions $Sa_1 \sim Sa_5$, and then

$$
Z = \sum_{i=1}^{5} \iint_{Sa_i} r B_{ri}^2(z)\mathrm{d}S
$$

$$
= \frac{1}{2}\left(D_{in}\Delta h + \frac{\Delta h^2}{4}\right)\Lambda_1 + \sum_{i=2}^{4}\left(D_{in}\Delta h + \frac{i\Delta h^2}{2}\right)\Lambda_i \quad (7)
$$

$$
+ \frac{1}{2}\left(D_{out}\Delta h - \frac{\Delta h^2}{4}\right)\Lambda_5
$$

where D_{in} and D_{out} denote inner and outer diameters of the metal tube respectively, Λ_i ($i = 1\sim5$) respectively denote calculation values of $\int B_{ri}^2(z)\mathrm{d}z$ which are obtained with the help of FE software.

If the inner and outer diameters of the metal tube vary, calculation of electromagnetic force on PM can be achieved only by choosing lines of different numbers and positions to analyze with variation of Sa in the surface integral.

4 EXPERIMENTS AND RESULTS

Experimental operating system mainly includes the coils fixed on iron stand (which energized before the PM falling, make the PM suspended in a fixed position), permanent magnet, vertical mental tube, and pulleys, as shown in Figure 7. Parameters of PM and mental tube are shown in Table 1. The facility for measurement of velocity is composed by the photoelectric rotary encoder (which is coaxially connected with the upper pulley), Digital Signal Processor (DSP) development board and oscilloscope. Encoder outputs the pulse signal based on the shaft speed; the DSP transfers the pulse signal to the speed digital signal; then the DAC chip converts the digital signal into the voltage signal recorded by the oscilloscope.

The following two experiments were designed to validate the theory.

1. Ensuring that the size and material of PM and metal tube are changeless, experiment 1 analyzes the relationship between the electromagnetic force and velocity of PM.
2. Ensuring that the size of PM and metal tube are changeless, experiment 2 studies the influence of metal resistivity ρ on electromagnetic force by selecting different materials of metal tubes.

4.1 Relation between velocity and electromagnetic force

Assuming that mg is the gravitational force acting on the PM, the acceleration of PM is a, I_{rotary} and R denote respectively the moment of inertia and radius of the pulleys, according to the kinematics equation, we have

$$
mg - F_e = \left(m + \frac{2I_{rotary}}{R^2}\right)a \quad (8)
$$

In order to measure the value of $2I_{rotary}/R^2$, removing the metal tube in experimental device, the PM doesn't endure the electromagnetic force, $F_e = 0$. We can acquire the acceleration $a = 5.461$ m/s². By Eq. 8, we can obtain $2I_{rotary}/R^2 = 0.0523$ kg with $m = 0.06583$ kg and g $= 9.8011$ m/s².

In order to validate the relation between the velocity and electromagnetic force on the PM, under the

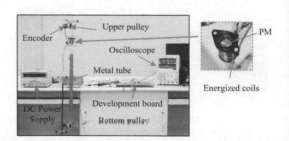

Figure 7. Experimental system.

86

condition that the materials and dimensions of PM and metal tube have no changes, velocity of the PM is measured surrounding a fixed position the metal tube. The results are shown in Figure 8.

Fitting the speed line recorded in the experiment, we can obtain relation between speed and time in form of a polynomial.

$$v = 9095t^3 - 179.1t^2 - 3.325t + 0.4376 \quad (9)$$

By taking the derivative of t in the polynomial in Eq. 9, we can get the acceleration a. According to Eq. 8, we can calculate the electromagnetic force F_e. From F_e/v curve in Figure 8, it can be seen that F_e/v keeps a constant whatever the velocity is, verifying the linear relation between speed and electromagnetic force of the PM surrounding a fixed position (i.e. Z is constant in Eq. 6).

4.2 Experiment of changing the resistivity

We make the PM falls through T2 copper tube, H62 brass tube and 6061 aluminum alloy tube with the same sizes in Table 1 respectively. The distance from the initial position of the center of the PM to the inlet of metal tube Δs keeps 0.1 m. The velocity curves measured are shown in Figure 9.

According to the experimental results shown in Figure 9, the whole movement process can be divided into five parts: ①Free falling process; ② Deceleration by electromagnetic force; ③Uniform motion of PM in the metal tube; ④Acceleration with the electromagnetic force reduced at the end of the tub; ⑤Free falling process.

In the three kinds of metal tubes, the acceleration in process ① and ⑤ are all the same, so it shows that at this process there exists no electromagnetic force. Inversely in the process ②, ③, ④ PM is under the effect of electromagnetic force. With the different resistivities of metal tube, the velocities during the processes of uniform motion and deceleration both change significantly.

The contrast between calculated and measured velocity is listed in Table 2. It can be seen from

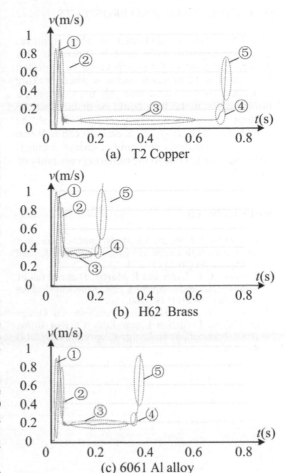

(a) T2 Copper

(b) H62 Brass

(c) 6061 Al alloy

Figure 9. The v–t curve with different metal resistivity.

Table 2. Calculated and measured velocity in uniform motion with different metal resistivity.

Materials of tube	Calculated velocity [m/s]	Measured velocity [m/s]	Relative error
T2 Copper	0.0941	0.0869	7.65%
H62 Brass	0.3864	0.3284	15.01%
6061 Al alloy	0.2176	0.1825	16.13%

Table 2 that theoretical calculation values and experimental results are almost the same. Due to the friction between pulleys and shafts and the fact that the PM cannot be ensured to move along the central axis precisely (which makes the friction between the PM and metal tube), the experimental measurements are smaller than the theoretical values.

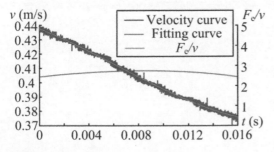

Figure 8. The velocity curve of PM.

5 CONCLUSIONS AND PROSPECTS

The PM's braking phenomenon when it moves in a nonmagnetic metal tube is studied in this paper. The results of research could be expected to be applied to practical occasion where braking is needed. Such as the commonly used elevator, a nonmagnetic metal tube could be installed at the bottom of the elevator shaft, and a PM could be installed at the bottom of the elevator cabin. When the cabin is suddenly dropping, a serious accident could be avoided due to the buffering capability of the system.

REFERENCES

M.H. Partovi, E.J. Morris, Electrodynamics of a magnet moving through a conducting pipe, Canadian Journal of Physics. 84 (2006) 253–271.

G. Donoso, C.L. Lader and P. Martin, Damped fall of magnets inside a conducting pipe, American Journal of Physics. 79(2011) 193–200.

B. Kou, Y. Jin, and H. Zhang. Analysis and Design of Hybrid Excitation Linear Eddy Current Brake, IEEE Transactions on Energy Conversions, 29(2014) 496–506.

K. Lee, C. Kim, and K. Park. Development of an Eddy-Current-Type Magnetic Floor Hinge, IEEE Transactions on Industrial Electronics, 53(2006) 561–568.

L.K. Wang, W.L. Li, F.Y. Huo, et al. Influence of Under-excitation Operation on Electromagnetic Loss in the End Metal Parts and Stator Step Packets of a Turbo-Generator, IEEE Transactions on Energy Conversions, 29 (2014) 748–757.

M. Horii, N. Takahashi and T. Narita. Investigation of Evolution Strategy and Optimization of Induction Heating Model, IEEE Transactions on Magnetics. 36(2000) 1085–1088.

I.D. Adewale, G.Y. Tian. Decoupling the Influence of Permeability and Conductivity in Pulsed Eddy-Current Measurements, IEEE Transactions on Magnetics, 49(2013) 1119–1127.

K. Shima, T. Fukami and K. Miyata, Analysis of eddy-current losses in solid iron under dc-biased magnetization considering minor hysteresis loops, Electrical Engineering in Japan. 188(2014) 56–66.

L.K. Wang, F.Y. Huo, W.L. Li, et al. Influence of Metal Screen Materials on 3-D Electromagnetic Field and Eddy Current Loss in the End Region of Turbo Generator. IEEE Transactions on Magnetics, 49(2013) 939–945.

Electrical, Control Engineering and Computer Science – Liu (Ed.)
© *2016 Taylor & Francis Group, London, ISBN 978-1-138-02937-8*

On-line dissolved gas analysis monitor based on laser photoacoustic spectroscopy

Jianbin Huang, Jianmin Tan & Long Di
Zhaoqing Power Supply Bureau, Guangdong Power Grid Corporation, Zhaoqing, China

ABSTRACT: Online DGA (Dissolved gas-in-oil analysis) is a sensitive and convincing technique for in-service detection of incipient faults of oil-immersed transformers. An On-line DGA monitor has been developed based on the Laser Photoacoustic Spectroscopy (LPAS) technique and has experimentally demonstrated great reliability and validity. It can analyze CH_4, C_2H_6, C_2H_4, C_2H_2, CO_2, CO, and H_2 in transformer insulted oil. In addition, the monitor has shown excellent stability, great sensitivity, and outstanding accuracy. All the relative detection errors of the target gases are smaller than 5%. The on-line DGA monitor would gain in popularity countrywide for the foreseeable future.

Keywords: oil-immersed transformer; Dissolved das-in-oil Analysis (DGA); on-line monitor; Laser Photoacoustic Spectroscopy (LPAS)

1 INTRODUCTION

Large power transformer is one of the major equipments of the power system and its failure can cause serious problems to the whole power system. The reliability of the power transformer is such significance. Thus, it is essential to monitor the health condition of transformers in real time. DGA is the widely used method for detection of incipient faults of oil-immersed transformers in service, since the concentrations of specific dissolved gas in insulation oil can be used to indicate the transformer's state. Hence, such equipment, which possesses the ability to on-line monitor dissolved gas concentrations in transformer insulation oil (called on-line monitor hereafter), could play a significant role in power system.

In laboratory, DGA used gas chromatograph and other electrochemical sensors. Hence, the first edition of the on-line monitor equipment originated from the gas chromatograph. By decreasing the size of the laboratorial gas chromatograph and redesigning some pipelines, it can be directly used for online monitor. In addition, this kind of online monitor has been commonly used in different sizes of transmission and distribution substations in China nowadays. However, the on-line monitor based on gas chromatograph has various problems. For example, it needs carrier gas and standard gas to operate, which leads to an installation of high pressure tanks beside and thus results in potential safety hazards.

Laser Photoacoustic Spectrum (LPAS) is a new technique. The application of on-line DGA equipment based on the LPAS technique can solve the problems presented in the utilizing of gas chromatograph. This article reports the development of an on-line DGA monitor based on LPAS technique, and the comparison of the gas concentrations measured by the equipment and a conventional gas chromatograph.

2 AN ON-LINE DGA MONITOR BASED ON LPAS TECHNIQUE

LPAS technique is based on the photoacoustic effect. When the gas is illuminated by laser, and the gas has absorbed the light energy of the corresponding wavelength, the gas molecules change its state, and through a rapid radiation process returns to its original state. When the gas molecules return to its original, it leads to the increase in the gases' temperature. Additionally with the laser illumination causing a change in the frequency of strength or wavelength modulation, the gas temperature will change proportionally in accordance with the frequency, and its temperature will also change resulting in a pressure change with a sound signal being generated, and thus this is gases' photo-acoustic effect.

Two main problems should be solved when applying LPAS technique to on-line DGA monitor. One, selecting appropriate characteristic wavelength in terms of each kind of gas and then choose suitable laser to irradiate. The other one is obtaining clear equation or relationship between the photoacoustic signal intensity and measured gas concentration.

Typically, seven gases exist among dissolved gas in transformer insulation oil. They are Methane (CH_4), Ethane (C_2H_6), Ethylene (C_2H_4), Acetylene (C_2H_2), Carbon dioxide (CO_2), Carbon monoxide (CO), and Hydrogen (H_2). This on-line DGA monitor adopts six semiconductor lasers to measure the concentrations of CH_4, C_2H_6, C_2H_4, C_2H_2, CO_2, and CO, respectively. The characteristic wavelengths of these six lasers are presented in Table 1. The photoacoustic signal intensity is proportional to the laser power, gas absorption constant, and gas concentration. Thus, the clear equation between photoacoustic signal intensity and gas concentration can be acquired as long as stabilizing laser power since gas absorption constant would not change.

The method of hydrogen measurement was different from other gases. The molecular weight of hydrogen is significantly smaller than other gases. The volume fraction of hydrogen can be obtained by measuring the variations of acoustic velocity, since the other gases' acoustic velocity's relative shift is proportional to the volume fraction of hydrogen.

With a light source emitting light at a specific frequency, seven kinds of fault gases irradiated with light absorption wavelength of the gas cell. Using high degree of fault gases microphone acoustic signal measured in the photoacoustic effect generated, the signal is filtered, since the correlation processing such as Fourier transform filters out the ambient noise, the signal stored in the embedded computer, for fault gas line analysis software for analysis and processing. The schematic of operation principle is shown in Figure 1.

Table 1. The characteristic wavelengths of six lasers.

Laser	Fault gas	Characteristic wavelength/nm
1	Methane (CH_4)	1651
2	Ethane (C_2H_6)	1679
3	Ethylene (C_2H_4)	1626
4	Acetylene (C_2H_2)	1520
5	Carbon dioxide (CO_2)	1580
6	Carbon monoxide (CO)	1567

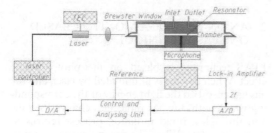

Figure 1. The schematic of operation principle.

3 SITE TEST AND RESULT

The on-line DGA monitor was installed in a substation to check the transformer reliability and validity. Every four months, the oil sample would be sent to laboratory and test by Gas Chromatograph (GC). These two analytic methods were conducted in the same temperature and humidity. The results are shown in Table 2.

The values of gas concentrations measured by these two methods of GC and LPAS in different dates were pretty close to Table 2. The values of CH_4, C_2H_6, C_2H_4, C_2H_2, CO_2, and CO measured by the method of LPAS were a little bit smaller than the method of GC. However, the values of the values of H2 had opposite trend.

The relative error was calculated as $|$ GC—LPAS $| / GC \times 100\%$. The relative errors of C_2H_6 and C_2H_4 were slightly larger than the others, but all the values were smaller than 5%. The result undoubtedly proved the reliability of the on-line DGA monitor based on LPAS technique.

Table 2. Gas concentrations in ppm measured by the methods of GC and LPAS in different dates.

Date	Method	CH^4	C_2H_6	C_2H_4	C_2H_2	CO_2	CO	H_2
2013.03.31	GC	5.35	4.37	86.66	3.43	9846.4	297.27	8.95
	LPAS	5.3	4.2	83.2	3.4	9733.2	295.6	9.2
2013.07.31	GC	4.93	4.09	81.48	4.07	7423.18	195.26	3.32
	LPAS	4.9	3.9	80.1	4.0	7312.4	195.1	3.4
2013.12.31	GC	5.85	5.44	119.98	2.91	10066.77	296.95	6.88
	LPAS	5.7	5.2	114.7	2.9	9914.55	295.3	7.0

4 CONCLUSION

This on-line DGA monitor is based on the LPAS technique, hence comprising a lot of advantages, such as no need for carrier gas and standard gas, not necessary to separate gas for measuring. These advantages eliminate the potential safety hazards and make the maintenance simple and easily. The results showed that the monitor owns excellent stability, great sensitivity and huge precision accuracy. Besides taking its relatively low cost into consideration, this self-developed online DGA equipment has good application foreground in China.

REFERENCES

Bo Liu, Feng Jiang. Online monitoring of dissolved gas in transformer insulation oil based on photoacoustic spectroscopy technique, East China Electric Power. 38 (2012) 250–253.

C.-P. Hung, M.-H. Wang, Diagnosis of incipient faults in power transformers using CMAC neural network approach, Electric Power Systems Research. 71 (2004) 235–244.

Huo-Ching Sun, Yann-Chang Huang, Chao-Ming Huang, A review of dissolved gas analysis in power transformers. Energy Procedia. 14 (2012) 1220–1225.

Jingsong Li, Xiaoming Gao, Weizheng Li, and so on, Near-infrared diode laser wavelength modulation-based photoacoustic spectrometer, Spectrochimica Acta Part A. 64 (2006) 338–342.

Khmais Bacha, Seifeddine Souahlia, Moncef Gossa, Power transformer fault diagnosis based on dissolved gas analysis by support vector machine, Electric Power Systems Research. 83 (2012) 73–79.

Michel, Duval. A review of faults detectable by gas-in-oil analysis in transformers, IEEE Electrical Insulation Magazine. 18 (2002) 8–17.

W.H. Tang, J.Y. Goulermas, Q.H. Wu, Z.J. Richardson, J. Fitch, A probabilistic classifier for transformer dissolved gas analysis with a particle swarm optimizer, IEEE Transactions on Power Delivery. 23 (2008) 751–759.

Zhiying Wu, Yinhai Gong, Qingxu Yu. Photoacoustic spectroscopy detection and extraction of discharge features gases in transformer oil based on 1.5 μ tunable fiber laser, Infrared Physics & Technology. 58 (2013) 86–90.

Electrical, Control Engineering and Computer Science – Liu (Ed.)
© 2016 Taylor & Francis Group, London, ISBN 978-1-138-02937-8

The third-order Absorbing Boundary Condition for finite difference modeling of first-order stress-velocity acoustic equation

Peng Song, Jun Tan & Dongming Xia
College of Marine Geo-Science, Ocean University of China, Qingdao, China
Key Laboratory of Submarine Geosciences and Prospecting Techniques, Ministry of Education, Qingdao, China

Jing Li, Zhaolun Liu & Bo Zhu
College of Marine Geo-Science, Ocean University of China, Qingdao, China

ABSTRACT: The third-order Absorbing Boundary Condition (ABC) is introduced into the forward modeling of first-order stress-velocity acoustic equation and the finite difference scheme of the third-order ABC is derived in the paper. The result of numerical test demonstrates that the third-order ABC has much higher absorbing efficiency than the conventional second-order ABC especially for the waves with large incident angles.

Keywords: third-order; Absorbing Boundary Condition; staggered-grid; forward modeling

1 INTRODUCTION

It is necessary for the seismic forward modeling to introduce artificial boundaries to define the computing area. However, an improper artificial boundary will produce false reflections that can reduce the accuracy of wave-field simulation seriously, so the artificial boundary processing has been an important part of the research of wave equation numerical modeling.

There are mainly three types of artificial boundary methods currently. The first one is the so called "damping layers" or "sponge layers" method (Israeli & Orszag 1981, Cerjan et al. 1985, Kosloff & Kosloff 1986), which introduces sponge zones near the boundary. The seismic waves propagating in the sponge zones can be attenuated exponentially. Unfortunately, it is generally very difficult to find a proper attenuation function to absorb incident waves perfectly (Liu & Sen 2010, 2012).

The second one is the Absorbing Boundary Condition (ABC) method. The conventional ABC method proposed by Enguist and Majda (1977, 1979), is based on the theory of approximation of one-way wave equation and derived through the Padé approximation for the one-way wave equation. Usually, the ABC equation derived through the high-order Padé approximation will include high-order partial derivatives, which are very difficult to be solved by numerical methods, so generally the second-order Padé approximation has to be used to derive the second-order ABC,

which, in fact, leads to poor absorption for the waves with large incident angles.

The third method is the Perfectly Matched Layer (PML) method (Berenger 1994, Hastings et al 1996, Chew & Liu 1996, Collino & Tsogka 2001, Hu et al. 2007, Gao & Zhang 2008). In theory, PML has excellent absorption for the seismic waves with arbitrary incident angles and arbitrary frequencies. However, in order to achieve excellent absorbing effect, dozens of matched layers need to be placed on the boundary, so PML is the most expensive computationally of the three methods (Liu & Sen 2010, 2012).

Higdon ABC (Higdon 1986, 1987, 1994) is another type of ABC method which is equivalent to the conventional ABC. Because the high-order Higdon ABCs involve only one-dimensional spatial derivatives, they are easier to be solved by finite difference method than the conventional high-order ABCs. Hence, the third-order Higdon ABC is introduced into the simulation of first-order stress-velocity acoustic equation in the paper. Model test shows that the third-order ABC has more excellent absorption than the second-order ABC especially for the waves with large incident angles, and the extra memory consumption and calculation amount are not too large.

In the next section, we review the numerical simulation of first-order stress-velocity acoustic equation. In section 3, we derive the finite difference scheme of the third-order ABC equation in detail. In section 4, the absorbing effects of the third-order ABC and the second-order ABC are compared by one model test. We end with some concluding remarks in section 5.

2 NUMERICAL SIMULATION OF FIRST-ORDER STRESS-VELOCITY ACOUSTIC EQUATION

Two-dimensional acoustic equation can be represented by the first-order stress-velocity equation generally (Mu & Pei 2005):

$$\frac{\partial v_x}{\partial t} = -\frac{1}{\rho}\frac{\partial p}{\partial x},$$

$$\frac{\partial v_z}{\partial t} = -\frac{1}{\rho}\frac{\partial p}{\partial z}, \tag{1}$$

$$\frac{\partial p}{\partial t} = -\rho V^2 \left(\frac{\partial v_x}{\partial x} + \frac{\partial v_z}{\partial z}\right)$$

where p is the stress; v_x and v_z represent the particle velocity components in the horizontal direction and the vertical direction respectively; ρ is the medium density; V is the velocity of seismic wave propagation. When $t = 0$, p, v_x and v_z are all zero except at the location of the source. The staggered-grid finite difference method is often used to solve the above equations. The staggered-grid finite difference scheme with second-order accuracy in time and $2L$th-order (L is a natural number) accuracy in space is given as follows:

$$v_{x\,i,j}^k = v_{x\,i,j}^{k-1} - \frac{1}{\rho}\frac{\Delta t}{\Delta x}\sum_{m=1}^{L} a_m(p_{i+m,j}^{k-1} - p_{i-m+1,j}^{k-1}),$$

$$v_{z\,i,j}^k = v_{z\,i,j}^{k-1} - \frac{1}{\rho}\frac{\Delta t}{\Delta z}\sum_{m=1}^{L} a_m(p_{i,j+m}^{k-1} - p_{i,j-m+1}^{k-1}),$$

$$p_{i,j}^k = p_{i,j}^{k-1} - \rho V^2 \frac{\Delta t}{\Delta x}\sum_{m=1}^{L} a_m(v_{x\,i-1+m,j}^k - v_{x\,i-m,j}^k) \tag{2}$$

$$-\rho V^2 \frac{\Delta t}{\Delta z}\sum_{m=1}^{L} a_m(v_{z\,i,j-1+m}^k - v_{z\,i,j-m}^k)$$

where Δx and Δz represent the space intervals in the horizontal direction and the vertical direction respectively; Δt is the time sampling interval; α_m is the difference coefficient. The superscripts and the subscripts stand for the time sampling point and the space grid point respectively.

When Equation 2 is applied in the model region of interest for the numerical simulation, some artificial boundary methods are needed on the boundary. If the sponge layers method or the PML method is used, dozens of layers need to be placed on the boundary, but for the third-order ABC, only L (L is less than 6 usually) layers are needed, so the memory consumption and calculation amount of

the sponge layers method and the PML method are much larger than those of the third-order ABC.

3 THE THIRD-ORDER HIGDON ABC EQUATION AND ITS DIFFERENCE SCHEMES

According to the research results of Higdon (1986), the third-order Higdon ABC equation can be written as follows (taking the left ABC for example):

$$B_3 U = \left[\left(\cos\alpha_1\frac{\partial}{\partial t} - v\frac{\partial}{\partial x} + \varepsilon\right)\left(\cos\alpha_2\frac{\partial}{\partial t} - v\frac{\partial}{\partial x} + \varepsilon\right)\right.$$
$$\left.\left(\cos\alpha_3\frac{\partial}{\partial t} - v\frac{\partial}{\partial x} + \varepsilon\right)\right]U = 0 \tag{3}$$

where B_3 is the operator of the third-order ABC; U represents any variable ($U = (p, v_x, v_z)$); v is the wave velocity; α_j is the angle parameter; ε is the damping factor (which is a very small positive number).

To derive the difference scheme of the third-order ABC, three operators, I, K, Z, are introduced, which are defined by $IU_{ij}^k = U_{ij}^k$, $K^n U_{ij}^k = U_{i+n,j}^k$, $Z^n U_{ij}^k = U_{ij}^{k+n}$. The difference schemes of three first-order boundary operators ($\cos\alpha_j (\partial/\partial t) - v(\partial/\partial x) + \varepsilon, j = 1, 2, 3$) in the third-order ABC equation can be expressed as:

$$G_j = l_{j,1}I + l_{j,2}K + l_{j,3}Z^{-1} + l_{j,4}KZ^{-1} \ (j = 1, 2, 3) \tag{4}$$

where

$$l_{j,1} = \Delta x\cos\alpha_j + v\Delta t + 2\Delta x\Delta t\varepsilon,$$
$$l_{j,2} = \Delta x\cos\alpha_j - v\Delta t,$$
$$l_{j,3} = -\Delta x\cos\alpha_j + v\Delta t, \tag{5}$$
$$l_{j,4} = -\Delta x\cos\alpha_j - v\Delta t$$

So the difference scheme of Equation 3 can be written:

$$G_1 G_2 G_3 U_{i,j}^{k+1} = 0 \tag{6}$$

The third-order boundary difference operator $G_1 G_2 G_3$ can be obtained by multiplication of the second-order boundary difference operator $G_1 G_2$ and the first-order boundary difference operator G_3, so $G_1 G_2$ must be calculated firstly.

$G_1 G_2$ can be derived by the multiplication of G_1 and G_2:

$$G_1 G_2 = s_{1,2}^1 I + s_{1,2}^2 K^2 + s_{1,2}^3 Z^{-2} + s_{1,2}^4 K^2 Z^{-2} + s_{1,2}^5 K$$
$$+ s_{1,2}^6 Z^{-1} + s_{1,2}^7 K^2 Z^{-1} + s_{1,2}^8 KZ^{-2} + s_{1,2}^9 KZ^{-1} \tag{7}$$

where $s_{1,2}^j$ ($j = 1, 2 \ldots 9$) are the difference coefficients which can be obtained by substituting Equation 5 to Equation 7.

The third-order boundary difference operator can be obtained by multiplying G_1, G_2 and G_3, which is expressed as:

$$
\begin{aligned}
G_1 G_2 G_3 = {} & r^1 I + r^2 K^3 + r^3 Z^{-3} + r^4 K^3 Z^{-3} + r^5 K \\
& + r^6 Z^{-1} + r^7 K^2 + r^8 Z^{-2} + r^9 K^3 Z^{-1} \\
& + r^{10} K Z^{-3} + r^{11} K^3 Z^{-2} + r^{12} K^2 Z^{-3} \\
& + r^{13} K Z^{-1} + r^{14} K^2 Z^{-1} + r^{15} K Z^{-2} \\
& + r^{16} K^2 Z^{-2}
\end{aligned}
\tag{8}
$$

where r^j ($j = 1, 2 \ldots 16$) can be obtained by substituting Equation 5 to Equation 8.

Then Equation 6 can be written:

$$
\begin{aligned}
(& r^1 I + r^2 K^3 + r^3 Z^{-3} + r^4 K^3 Z^{-3} + r^5 K \\
& + r^6 Z^{-1} + r^7 K^2 + r^8 Z^{-2} + r^9 K^3 Z^{-1} + r^{10} K Z^{-3} \\
& + r^{11} K^3 Z^{-2} + r^{12} K^2 Z^{-3} + r^{13} K Z^{-1} + r^{14} K^2 Z^{-1} \\
& + r^{15} K Z^{-2} + r^{16} K^2 Z^{-2}) U_{i,j}^{k+1} = 0
\end{aligned}
\tag{9}
$$

According to Equation 9, the difference scheme of the ith layer boundary can be obtained:

$$
\begin{aligned}
U_{i,j}^{k+1} = (& r^2 U_{i+3,j}^{k+1} + r^3 U_{i,j}^{k-2} + r^4 U_{i+3,j}^{k-2} + r^5 U_{i+1,j}^{k+1} \\
& + r^6 U_{i,j}^{k} + r^7 U_{i+2,j}^{k+1} + r^8 U_{i,j}^{k-1} + r^9 U_{i+3,j}^{k} \\
& + r^{10} U_{i+1,j}^{k-2} + r^{11} U_{i+3,j}^{k-1} + r^{12} U_{i+2,j}^{k-2} + r^{13} U_{i+1,j}^{k} \\
& + r^{14} U_{i+2,j}^{k} + r^{15} U_{i+1,j}^{k-1} + r^{16} U_{i+2,j}^{k-1}) / (-r^1)
\end{aligned}
\tag{10}
$$

The difference schemes of the third-order ABC for the right, top and bottom boundary can be derived in the same way.

In addition, it is noteworthy that the Higdon ABC is based on the separation of the in-going and the out-going waves. If the frequency and the wavenumber are zero or in the near of zero (when the error exceeds the accuracy range of the computer), it will be difficult to separate the in-going and the out-going waves completely. When such a signal mixed with low frequency and wavenumber waves arrives at the boundary, some in-going waves with low frequency and wavenumber will be reflected into the central area and be amplified with the growth of computing time, which will cause serious instability in long seismic records. The instability appears only in high-order ABCs (Higdon 1987), and the higher the order is, the more serious the instability will be.

Two ways are suggested to overcome the instability of the third-order ABC in the paper.

One is trying to use the double precision calculation and the other is to choose a proper damping factor ε in Equation 3. We suggest that if double precision is used, the instability of the third-order ABC can be overcome by setting ε as $0.001 \sim 0.005$.

4 NUMERICAL EXAMPLE

In the paper, a zero phase Ricker wavelet with the main frequency of 30 Hz is used as a source function and the difference scheme with second-order accuracy in time and 8th-order accuracy in space is applied to compute the center wave-field.

A model with constant velocity and density (the velocity is 4000 m/s and the density is 2000 kg/m³) is applied to test the absorption of the third-order ABC. The horizontal distance and depth of the model are both 3750 m. The horizontal and vertical space intervals are both 5 m and the time sampling

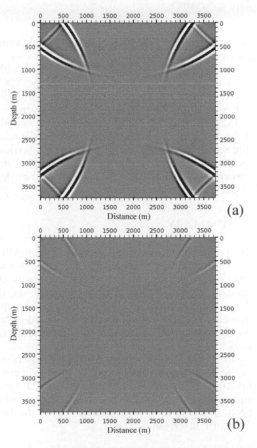

Figure 1. Snapshots at 0.85 s with the second-order ABC (a), and the third-order ABC (b).

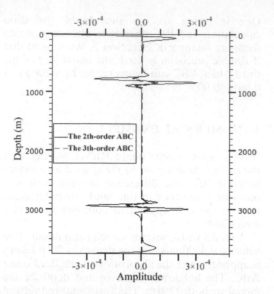

Figure 2. The comparison of the two single traces in Figure 1a and Figure 1b at Distance = 450 m ("—" represents the second-order ABC and "– –" represents the third-order ABC).

interval is 0.5 ms. The shot point is located in the center of the model.

Figure 1 shows the snapshots at 0.85 s with the second-order ABC and the third-order ABC. Figure 2 shows the comparison of the two single traces in Figure 1a and Figure 1b at Distance = 450 m.

Figure 1 and Figure 2 illustrate that the second-order ABC has high absorption only for the seismic waves with vertical incidences or nearly vertical incidences, while the third-order ABC has excellent absorption no matter whether the incident angles of the waves are small or large.

5 SUMMARY

The paper introduces the third-order Higdon ABC into the boundary processing for the staggered-grid numerical modeling of first-order stress-velocity acoustic equation. Numerical experiment shows that the third-order ABC has much higher absorption efficiency than the conventional second-order ABC especially for the waves with large incident angles.

Moreover, the third-order ABC can be extended to the boundary processing of the wave-field continuation of elastic wave equations or 3D wave equations, and it also has broad application prospects in the reverse-time migration and the full waveform inversion.

REFERENCES

Berenger, J.P., "A perfectly matched layer for the absorption of electromagnetic waves," Journal of Computational Physics, Vol. 114, No. 2, 1994, pp. 185–200.

Cerjan, C., R. Kosloff, and M. Reshef, "A nonreflecting boundary condition for discrete acoustic and elastic wave equations," Geophysics, Vol. 50, No. 4, 1985, pp. 705–708.

Chew, W.C., and Q.H. Liu, "Perfectly matched layers for elastodynamics: A new absorbing boundary condition," Journal of Computational Acoustics, Vol. 4, No. 4, 1996, pp. 341–359.

Collino, F., and C. Tsogka, "Application of the perfectly matched absorbing layer model to the linear elastodynamic problem in anisotropic heterogeneous media," Geophysics, Vol. 66, No. 1, 2001, pp. 294–307.

Enquist, B., and A. Madja, "Absorbing boundary conditions for the numerical simulation of waves," Math. Comp., Vol. 31, No. 139, 1977, pp. 629–651.

Enquist, B., and A. Madja, "Radiation boundary conditions for acoustic and elastic wave calculations," Communications on Pure and Applied Mathematics, Vol. 32, No. 3, 1979, pp. 313–357.

Gao, H., and J. Zhang, "Implementation of perfectly matched layers in an arbitrary geometrical boundary for elastic wave modeling," Geophysical Journal International, Vol. 174, No. 3, 2008, pp. 1029–1036.

Hastings, F.D., Schneider, J.B., and Broschat, S.L., "Application of the Perfectly Matched Layer (PML) absorbing boundary condition to elastic wave propagation," The Journal of the Acoustical Society of America, Vol. 100, No. 5, 1996, pp. 3061–3069.

Higdon, R.L., "Absorbing boundary conditions for difference approximations to the multidimensional wave equation," Mathematics of computation, Vol. 47, No. 176, 1986, pp. 437–459.

Higdon, R.L., "Numerical absorbing boundary conditions for the wave equation," Mathematics of computation, Vol. 49, No. 179, 1987, pp. 65–90.

Higdon, R.L., "Radiation boundary conditions for dispersive waves," SIAM Journal on Numerical Analysis, Vol. 31, No. 1, 1994, pp. 64–100.

Hu, W., A. Abubakar, and T.M. Habashy, "Application of the nearly perfectly matched layer in acoustic wave modeling," Geophysics, Vol. 72, No. 5, 2007, pp. SM169–SM175.

Israeli, M., and S.A. Orszag, "Approximation of radiation boundary conditions," J.comput. phys., Vol. 41, No. 1, 1981, pp. 115–135.

Kosloff, R., and D. Kosloff, "Absorbing boundaries for wave propagation problems," J. computat. phys., Vol. 63, No. 2, 1986, pp. 363–376.

Liu, Y., and M.K. Sen, "A hybrid scheme for absorbing edge reflections in numerical modeling of wave propagation," Geophysics, Vol. 75, No. 2, 2010, pp. A1–A6.

Liu, Y., and M.K. Sen, "A hybrid absorbing boundary condition for elastic staggered-grid modeling," Geophysics Prospecting, Vol. 60, No. 6, 2012, pp. 1114–1132.

Mu, Y.G., and Z.L. Pei, "Seismic numerical modeling for 3-D complex media," Petroleum Industry Press, Beijing, 2005.

Electrical, Control Engineering and Computer Science – Liu (Ed.)
© 2016 Taylor & Francis Group, London, ISBN 978-1-138-02937-8

Automatic vertical parking steering control based on fuzzy control and Extended Kalman Filter

Xuewu Ji, Si Sun & Yong Huang
State Key Laboratory of Automotive Safety and Energy, Tsinghua University, Beijing, P.R. China

ABSTRACT: A new kind of control strategy for automatic vertical parking has been investigated in this article. A kinematic model of the car in reverse process at low-speeds is built and analyzed. According to the parameters of experimental vehicle and collision avoidance constraints, the initial limited position for automatic vertical parking is reversely located. Combining drivers experience and the fuzzy control theory, a steering controller for automatic vertical parking is proposed. In order to better solve error accumulation problems caused by the noise of wheel speed sensor during actual automatic vertical parking, the kinematic model is optimized by Extended Kalman Filter (EKF). An optimal kinematic model and a fuzzy steering controller are built in Matlab/Simulink for the simulation of vertical parking process. The results show that, the fuzzy control works well on automatic vertical parking, and the EKF has good adaptability with solving error accumulation in parking system.

Keywords: automatic vertical parking; fuzzy control; Extended Kalman Filter

1 INTRODUCTION

With the development of urbanization in China, the number of cars is increasing every year, and parking environment is more and more complex. As for many drivers, parking is a time-consuming task. This enables many research institutions study automatic parking systems. Under the restrictions of collision avoidance in vertical parking conditions, Li planned parking path and fitted with B Spline, but no processes were mentioned; based on the fuzzy logic control and human experience, Lee respectively researched parking systems, while it needs a great number of experiments; applying genetic algorithms, Meng optimized the three parking steering control strategies, but vehicle speed fluctuation is not taken into consideration. In this paper, under the condition of vertical parking, considering the impact of speed fluctuations, applying the EFK algorithm and fuzzy control theory, automatic vertical parking conditions is studied.

2 KINEMATIC MODEL OF CAR STEERING AT LOW SPEED

The global coordinate system xoy is illustrated in Figure 1, picking the car's rear axle center as the reference point. For the low-speed car, lateral forces can be ignored on straight roads. No tire

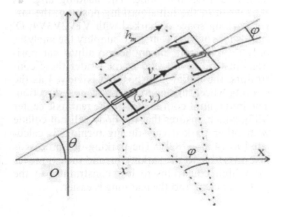

Figure 1. Vertical parking coordinate.

slips occurring, the axial velocity of rear center is approximately zero, then the derived vehicle kinematic model is as follows:

$$\begin{cases} \dot{x}_r = \cos\theta \cdot v_r \\ \dot{y}_r = \sin\theta \cdot v_r \\ \dot{\theta} = \dfrac{\tan\varphi}{h_m} \cdot v_r \end{cases} \qquad (1)$$

where, φ is the Ackerman angle (Front axle center angle), θ the heading angle, h_m the wheelbase,

(x_r, y_r) the coordinates of the rear axle center, and v_r the speed of the rear axle center.

Through the measuring disk of wheel angle and angle sensors, Ackerman angle φ and the wheel angle δ can be respectively collected. In Matlab, we choose LS fitting Equation 2 (angle unit: (°)), then the steering transmission ration $K = 20.16$:

$$\varphi \approx 0.0496\delta + 0.915 \tag{2}$$

3 DESIGN OF FUZZY CONTROLLER

3.1 Vertical parking environment

The surrounding environment is one of the key factors to the path planning. The size of the parking space affects directly the process. According to requirements for parking lot, vertical parking parameters are determined as 5.8 m long and 2.6 m wide. After the system finds the right parking space, steering wheel angle can be automatically regulated; the driver is in charge of speeds, to make sure that the vehicle will be parking into the target space.

Taking the case shown in Figure 2 as an example, the car is simplified to a rectangle 4382 mm long and 1835 mm wide. The heading angle is zero when in the initial parking position; the target parking space is marked with V1V2V3V4; O' stands for rotation center. To simplify the analysis, only reverse gears are applied to adjust car positions instead of forward gears. Under these constraints, the initial limit position is viewed as the steering wheel turning to the right-most direction. The initial limit coordinates of the rear axle center M[3], which ensures that the body will not collide with other parked cars during turning, is calculated as (4964, 7828). The parking-out process is just invertible to the parking-in, and the car moves from high constrains to low constraints, so the paths are clear and the planning is easier.

3.2 Design of fuzzy controller

The framework of vertical parking fuzzy control is illustrated in Figure 3. The inputs of kinematics model include steering angle φ of the front axle and the car speed v. And the angle φ is also the output of fuzzy controller. The steering wheel angle can be obtained in real time by φ times steering ratio K. The design steps are as follows.

① Fuzzification. As shown in Figure 4, the membership functions include the rear axle center (x, y), heading angle θ, and front steering angle φ. The counterclockwise is positive direction for θ, and the clockwise is positive direction.

② Establishing rules. According to the driver's operation experience, the input and output data are recorded during the driver's control. Fuzzy control rules are presented in Table 1.

③ Synthesis and Reasoning. The membership values can be synthesized from the 26 rules.

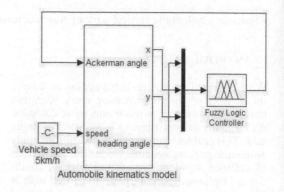

Figure 3. Vertical parking fuzzy control model.

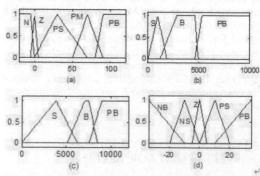

(a) Membership function of θ, (b) Membership function of x, (c) Membership function of y, (d) Membership function of φ.

Figure 4. Membership function.

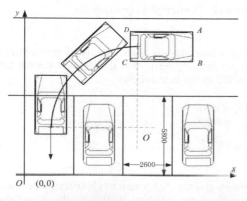

Figure 2. The initial limit position.

Table 1. Fuzzy control rules.

φ	x/y	S	B	PB
$\theta = N$	S	/	/	/
	B	/	PB	PB
	PB	/	PB	PS
$\theta = Z$	S	/	/	/
	B	/	PB	PB
	PB	/	PS	PS
$\theta = PS$	S	PB	PB	/
	B	PB	PS	PS
	PB	/	PS	Z
$\theta = PM$	S	PS	PB	/
	B	PS	PS	Z
	PB	/	NS	NB
$\theta = PB$	S	Z	Z	/
	B	NB	NS	/
	PB	/	/	/

Optimization of Kinematic Model.

④ Defuzzification. The output can be precisely achieved by gravity method.

4 OPTIMIZATION OF KINEMATIC MODEL BASED ON EKF

4.1 Shortage of the original kinematic model

When Kinematic Model Equation 1 is discrete, position at moment k satisfies Equation 3. If input speeds and the front steering angle at unit sampling time is Δt, then x, y and θ can be achieved.

$$\begin{cases} x_k = x_{k-1} + \triangle x_k = x_{k-1} + \cos\theta_k \cdot v_k \cdot \Delta t \\ y_k = y_{k-1} + \triangle y_k = y_{k-1} + \sin\theta_k \cdot v_k \cdot \Delta t \\ \theta_k = \theta_{k-1} + \triangle\theta_k = \theta_{k-1} + \dfrac{\tan\varphi_k}{h_m} \cdot v_k \cdot \Delta t \end{cases} \quad (3)$$

In real parking system, speeds and steering wheel angles are measured by sensors, inevitably along with some noise, which contributes to affect the control accuracy after integral amplification. To solve this problem, Extended Kalman Filter (EKF) is applied to optimize the estimation of parking position, to make it more adaptable.

4.2 Design of Kalman Filter

In the condition that the system structure is known, given the observation of moment k, the optimal solution can be solved through Kalman Filter (KF), which minimizes the estimation and covariance. As shown in Figure 5, the five formulas consist of the main points of the KF.

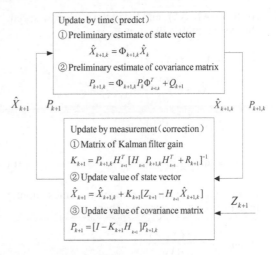

Figure 5. KF algorithm structure.

The calculation process of Kalman Filter is an iterative process, including time and measurement update. The filtering can be executive after the system matrix $\Phi_{k+1,k}$, measurement matrix H_{k+1}, and initial values of $\hat{X}_0$, P_0 are determined. It should be noted that Kalman Filter is only applicable to linear system. Extended Kalman Filter (EKF) is generally applied when it comes to nonlinear equations. The basic idea is to extend the nonlinear equations to the first order, truncating high orders, transforming to linear system. In this paper, EKF are applied in model Equation 1.

4.3 Optimization kinematic model based on EKF

First, the system matrix $\Phi_{k+1,k}$ and measurement matrix H_{k+1} should be determined. The system state vector is $X = [x \quad y \quad \theta \quad k_l \quad k_r]^T$, and measurement vector is $Z = [v \quad w]^T$. At the moment k, with the help of the two wheel speeds and the steering wheel angle, following can be achieved:

$$v_k = \frac{\pi R(k_l n_{1k} + k_r n_{2k})}{N\Delta t} \quad (4)$$

$$w_k = \frac{\pi R(k_l n_{1k} + k_r n_{2k})\tan\varphi_k}{N\Delta t h_m} \quad (5)$$

where, k_l and k_r are the coefficients for variation of the two tire radius (set 1 for simulation), n_{1k} and n_{2k} are the number of pulses of the rear wheels within one sampling period, N is the number of pulses for one revolution (40 for the experimental car JOYEAR X5), and w_k is the car's yaw speed. The system matrix $\Phi_{k+1,k}$ is:

$$\hat{X}_{k+1,k} = \hat{X}_k + f(X_k, k)$$

$$= \hat{X}_k + \frac{\partial f(X_k,k)}{\partial X_k}\Big|_{X_k=\hat{X}_k} \hat{X}_k$$

$$= \hat{X}_k\left(I + \frac{\partial f(X_k,k)}{\partial X_k}\right)$$

$$= \Phi_{k+1,k}\hat{X}_k \tag{6}$$

where, $f(X_k,k)$ is the increment in a unit sampling time. For the linear process, then:

$$\Phi_{k+1,k} = I + \frac{\partial f(X_k,k)}{\partial X_k}$$

$$=\begin{bmatrix} 1 & 0 & -\pi R(k_l n_{1k} + k_r n_{2k})\sin\theta_k/N \\ 0 & 1 & \pi R(k_l n_{1k} + k_r n_{2k})\cos\theta_k/N \\ 0 & 0 & 1 \\ 0 & 0 & 0 \\ 0 & 0 & 0 \end{bmatrix}$$

$$\begin{bmatrix} \pi R n_{1k}\cos\theta_k/N & \pi R n_{2k}\cos\theta_k/N \\ \pi R n_{1k}\sin\theta_k/N & \pi R n_{2k}\sin\theta_k/N \\ \pi R n_{1k}\tan\varphi_k/Nh_m & \pi R n_{2k}\tan\varphi_k/Nh_m \\ 1 & 0 \\ 0 & 1 \end{bmatrix} \tag{7}$$

The measurement vector is:

$$Z_k = \begin{bmatrix} v_k \\ w_k \end{bmatrix} = \begin{bmatrix} \dfrac{\pi R(k_l n_{1k} + k_r n_{2k})}{N\Delta t} \\ \dfrac{\pi R(k_l n_{1k} + k_r n_{2k})\tan\varphi_k}{N\Delta t h_m} \end{bmatrix} \tag{8}$$

Z_k is linearized as:

$$H_k = \frac{\partial Z_k}{\partial X_k}\Big|_{X_k=\hat{X}_k}$$

$$= \begin{bmatrix} 0 & 0 & 0 & \pi R n_{1k}/N\Delta t \\ 0 & 0 & 0\pi R n_{1k}\tan\varphi_k/N\Delta t h_m \end{bmatrix}$$

$$\begin{bmatrix} \pi R n_{2k}/N\Delta t \\ \pi R n_{2k}\tan\varphi_k/N\Delta t h_m \end{bmatrix} \tag{9}$$

After optimization, the kinematic model in Figure 3 is transformed to a model as shown in Figure 6. The EKF algorithm integrates

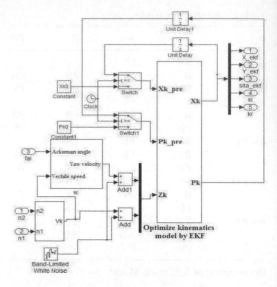

Figure 6. EKF algorithm structure.

independent signals from different sensors (the wheel speed and the steering wheel angle), and improves the position estimation accuracy by filtering redundancy.

5 SIMULATION RESULTS

To verify the fuzzy controller, simulations are carried out under the condition where v is a constant and φ is the direct output value of the fuzzy controller. The data used for experiment are shown in Table 2.

Reverse parking start position is set where the rear axle center is at the limit coordinates (4382, 1835). $v = 5$ km/h, vertical parking lot length: 5.8 m, width: 2.6 m. It takes 7.9 s for the parking. The simulation results are shown in Figure 7, which indicate that the fuzzy controller of vertical parking manages the whole parking process successfully. Figure 8 shows the heading angles changing during the parking process. When parking is completed, the car is 0.9 m away from the bottom of the space, and 0.4 m away from the left and right limits; the heading angle changes from 0° to 90°. The steering wheel turns to neutral position while the car is into the parking space. During whole parking process, no collisions with any other cars occur. The fuzzy control for vertical parking has good adaptability.

However, in the real parking process, it is hard to get desirable inputs. To make the simulation more realistic to the actual situation, white noise

Table 2. Vehicle parameters.

Parameters	Value
Length L_a	4382 mm
Wide W_a	1835 mm
Front track W_{bf}	1540 mm
Rear track W_{br}	1545 mm
Wheel base h_m	2690 mm
Front overhang L_f	947 mm
Rear overhang L_r	745 mm
Overall ratio K	20.16
Minimum turning radius R_{min}	4180 mm

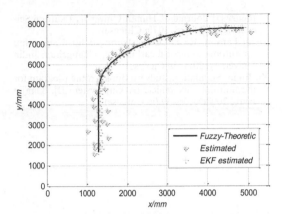

Figure 9. Track contrast before and after the optimization of the EKF.

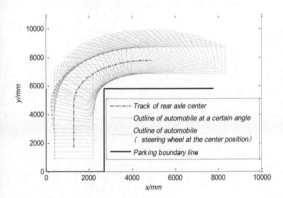

Figure 7. Trajectory of vertical parking.

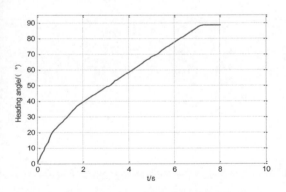

Figure 8. Heading angle of vertical parking.

(mean: 0, standard deviation: 0.7 km/h) is added into the wheel speed as disturbance.

In Figure 9, the black line indicates the trajectory without signal noise applied to the system, the blue dots suggest the discrete sampling trajectory with some noise, and the red dots show the trajectory after the optimization of EKF

with some noise. Comparing the ideal trajectory, the maximum deviations in x and y directions are respectively 5.1 cm and 9.6 cm. We can conclude that after the optimization to the kinematic model, the car's position is precisely estimated under noise disturbance, and its accurate estimation contributes to the development of automated parking systems.

6 CONCLUSION

In this article, a steering control strategy for vertical parking systems was proposed and the following conclusions could be drawn:

1. Fuzzy control theory for vertical parking had good adaptability, enabling car parking successfully.
2. EKF algorithm was implemented to optimize kinematic model. It effectively filtered the accumulation errors of sensors, making the simulation more practical.

REFERENCES

Design and research institute of the Beijing municipal government. CJJ37-90 Specification for Design of Municipal Roads [S]. Beijing: China Building Industry Press, 1998:130–133 (in Chinese).

Fu Mengyin. Kalman filtering theory and its application in the navigation system [M]. Beijing: Science Press, 2003:65–85 (in Chinese).

Lee J.Y, Lee J.J. Multiple designs of fuzzy control for car parking using evolutionary algorithm [C]. Proceedings of International Conference on Mechatronics. Kumamoto, Japan, 2007:1–6.

Li Hong, Guo Konghui, Song Xiaolin. Trajectory planning of automatic vertical parking based on spline theory [J]. Journal of Hunan University (Natural Sciences), 2012(7):25–30 (in Chinese).

Meng Fanwei. Research on steering control algorithm for automatic vertical parking [D]. Changchun: Jinlin University. 2011 (in Chinese).

Xu Jinjin. Simulation study on duplex route program for the autonomous vehicle parking system [J]. Tianjin Auto, 2008(5): 36–39 (in Chinese).

Zhang Cheng, Luo Yong, Chen Hui. Research of automatic parking pose estimation based on EKF algorithm [J]. Shanghai Auto, 2012(6):56–58, 62 (in Chinese).

Zhao Y. Collins Jr E.G. Robust automatic parallel parking in tight spaces via fuzzy logic [J]. Robotics and Autonomous Systems, 2005, 51(2): 111–127.

Electrical, Control Engineering and Computer Science – Liu (Ed.)
© 2016 Taylor & Francis Group, London, ISBN 978-1-138-02937-8

Fault diagnosis for transmission network based on Timing Bayesian Suspected Degree

Xiangfei Ma, Qing Chen & Zhanjun Gao
*Key Laboratory of Power System Intelligent Dispatch and Control (Shandong University),
Ministry of Education, Jinan, Shandong, China*

ABSTRACT: Errors and incompleteness may exist in the alarm information when fault occurs on transmission line [1]. To improve the accuracy and fault toleration of fault diagnosis, this paper presents a fault diagnosis algorithm for transmission network based on Timing Bayesian Suspected Degree. The algorithm models fault process in the form of probability using information from circuit breakers, then the ratio of actual fault symptoms to suspected fault symptoms reflects the importance of the actual fault symptoms. In addition, considering the timing characteristics of the alarm information [2], time-based symptom weight is introduced in the Bayesian Suspected Degree index to enhance the accuracy of fault diagnosis in uncertain situations.

Keywords: Bayesian Suspected Degree; fault diagnosis; time-based symptom weight

1 INTRODUCTION

Generally, the grid fault diagnosis means reasoning the fault process backward using the relay action and breaker tripping information to identify the primary fault. Common intelligent algorithms are expert system, Petri net, analytical model, artificial neural network and so on. However, these intelligent algorithms are more or less flawed, such as (1) models too complicated (analytical model), (2) slow convergence rate and learning speed (artificial neural network), (3) undesirable diagnostic efficiency (expert system), (4) poor fault-tolerance capability.

A fault diagnosis algorithm is proposed in this paper. The algorithm uses faulty component as a unit rather than circuit breakers in the calculation process, which reduces the dimensions of topological matrix and priori probability matrix, therefore reduces the computational complexity. A desirable diagnosis result can be obtained under complex network environment with timing characteristics of alarm information weighted.

2 THE BASIC MODEL

To analyze the relationship between fault and the alarm information, fault propagation process usually requires modeling. Commonly used methods contain Petri net, Bayesian network, causal diagram, etc. Considered to be the simplification of these models, bipartite graph model [3] retains the modeling capabilities of those foresaid models and simultaneously reduces the complexity of the algorithm, which has desirable application prospects.

We use probability weighted bipartite causal map as a fault propagation model, as shown in Figure 1. Three fault nodes are denoted by f while six symptoms are denoted by s. P is the probability of s when f has occurred. A priori probability matrix $P_{s|f}$ ($l \times m$) can be obtained from Figure 1, in which l indicates the number of faulty nodes while m indicates the number of symptoms. Based on the Bayesian formula (1), posterior probability matrix $P_{f|s}$ can be deduced from $P_{s|f}$. Posterior probability $P_{f|s}$ is the reflection of backward reasoning, which represents the fault probability when the symptoms have occurred.

$$p(f|s) = \frac{p(f)p(s|f)}{\sum_{f_i \in F} p(f_i)p(s|f_i)} \tag{1}$$

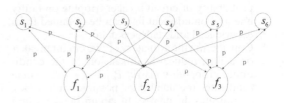

Figure 1. Bipartite graph model.

$$bsd_f = \frac{\sum\limits_{s \in S_N} p(f|s)}{\sum\limits_{s_j \in S} p(f|s_j)} \tag{2}$$

$$tbsd_f = \frac{\sum\limits_{s_i \in S_N} h_i p(f|s_i)}{\sum\limits_{s_j \in S} p(f|s_j)} \tag{3}$$

The ultimate goal of fault diagnosis is to identify the most likely fault node from three possible fault nodes under the condition of occurred symptoms. As a result, the explanatory ability of occurred symptoms to every fault node should be computed. Respectively calculate Bayesian Suspected Degree (*bsd*) of three nodes by formula (2), where S_N indicates actual symptom set while S indicates the possible symptom set of every fault node. The node whose *bsd* index is large means the occurred symptoms gives this fault node a good explanation.

Put all fault nodes in a descending order. If the first *k* nodes can cover all the occurred symptoms, we determine the first *k* nodes are faulty.

3 FAULT DIAGNOSIS ALGORITHM

The most important information obtained from dispatching center when fault occurs is tripped circuit breaker information. Due to the configuration of backup protection, the mapping relationship between fault and the symptoms is many-to-many, which means that the fault of one transmission line may cause numerous circuit breakers tripped on different lines, while one circuit breaker trip also may be caused by several different lines. Thus, in the bipartite causal map, $P \in (0,1)$. The fault diagnosis algorithm process is as follows:

1. Identify the missing circuit breaker information corresponding to the main protection information and then group the information in chronological order.
2. Search topology to find the corresponding possible fault set based on the actual tripped circuit breaker information.
3. Plot the bipartite causal map, where the priori probability of circuit breaker tripping on faulty line and on adjacent line can be obtained from experienced and statistical data.
4. Considering the symmetry of two circuit breaker on a line, reduce the dimension of the conditional probability matrix $P_{s|f}$ to $l \times m/2$, where *l* indicates the number of possible fault nodes, *m* indicates the number of circuit breakers, and *m*/2 indicates the dimension of the topological matrix.

5. Based on the Bayesian formula (1), posterior probability matrix $P_{f|s}$ can be deduced.
6. Weighted the circuit breaker tripping information according to chronological order, calculate the *tbsd* index of every fault in the possible fault set as formula (3), where *h* indicates the timing weight of circuit breaker information. In the next analysis, the value *h* equals 0.5 for the circuit breakers whose tripping time are above the action time of those tripping circuit breakers controlled by the main protection.
7. Put all fault nodes in a descending order according to the *tbsd* index.
8. If the first n nodes can cover all the tripping circuit breakers, determine the first k nodes are faulty; otherwise, increase n in order until all the tripping circuit breakers can be covered.
9. Verify the correctness of the located faulty components. When two faulty lines are adjacent, the line with minor information redundancy is usually failed to identify. Guarantee the faulty components cover all the lines whose main protection acts.
10. Finally analyze the missing-information components and the mal-operation components.

4 EXAMPLE

In order to test the effectiveness of the method described herein, a 500 kV power wiring diagram is used as an example (see Fig. 2). In the figure, L indicates the transmission line, B indicates the bus and CB indicates the circuit breaker.

For example, L_1 is supposed to be faulty, the action information includes CB4, CB5, CB8 and

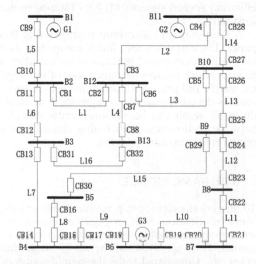

Figure 2.   The 500 kV transmission system.

104

Table 1. Diagnosis result when L_1 is faulty.

Action information	CB4, CB5, CB8; P_{L1m}, $P_{L2-B11s}$, $P_{L3-B10s}$, $P_{L4-B13s}$
Breakers	1, 4, 5, 8
Suspected faulty set	1, 2, 3, 4, 5, 6, 13, 14, 16
tbsd in descending order	0.4806, 0.1891, 0.1891, 0.1504, 0.0065, 0.0056, 0.0032, 0.0013, 0
Arranged faulty set	1, 4, 2, 3, 14, 5, 16, 13, 6
Fault diagnosis	L_1 is faulty; CB1 missed information; CB2 failed to operate
Time [s]	0.127539

Table 2. Fault diagnosis results on several fault scenarios.

No	Action information	Fault node	Time [s]	Result analysis
1	CB1, CB2; P_{L1m}	L_1	0.0977	Correct operation
2	CB1, CB4, CB5, CB8; P_{L1m}, $P_{L2-B11s}$, $P_{L3-B10s}$, $P_{L4-B13s}$	L_1	0.0939	CB2 failed to operate
3	CB2; P_{L1m}	L_1	0.0876	CB1 missed information
4	CB4, CB5, CB8; P_{L1m}, $P_{L2-B11s}$, $P_{L3-B10s}$, $P_{L4-B13s}$	L_1	0.1275	CB1 missed information; CB2 failed to operate
5	CB1, CB2, CB3, CB4; P_{L1m}, P_{L2m}	L_1, L_2	0.3726	Correct operation
6	CB1, CB2, CB15, CB16, CB19, CB20; P_{L1m}, P_{L8m}, P_{L10m}	L_1, L_8, L_{10}	0.1400	Correct operation

P_{L1m}, $P_{L2-B11s}$, $P_{L3-B10s}$. Firstly we identify the missing circuit breaker CB1. In chronological order, CB1 is grouped in the first order while CB4, CB5, CB8 are grouped in the second order. From all the breaker information, the original topological matrix (16×16) is simplified to a new matrix (9×16), where 9 means the number of suspected fault components. Then the posterior probability matrix $P_{f|s}$ (9×16) can be deduced. Calculate the *tbsd* index of every row of $P_{f|s}$, some simulation results are as Table 1.

Based on the Matlab platform, several fault scenarios is verified in this paper, such as mal-operation, missing information and complicated fault (Table 2). In the Action information column of Table 2, CB indicates the circuit breaker, P_{L1m} indicates the main protection of line 1 and $P_{L2-B11s}$ indicates the reserve protection near bus 11 of line 2. As can be seen from Table 2, the algorithm can give the correct results under these conditions; algorithms can also obtain the correct diagnosis on multiple lines of adjacent and non-adjacent. With location time less than 0.4 s, the algorithm can be used for online diagnosis.

proposed algorithm can deal with the incomplete and wrong alarm information. In the computing process, the maximum dimension of matrix is faulty component, substituting the mode of using circuit breaker as a diagnosis unit in traditional methods, which to some extent simplify the computing complexity. Considering the timing characteristics of alarm information, the algorithm enriches the dimension of fault section information. To conclude, the method proposed has a good prospect.

REFERENCES

Guo Chuangxin, Zhu Chuanbai, Cao yijia. "State of arts of fault diagnosis of power system". Automation of Electric Power Systems, 2006, 30(8): 98–103 (in Chinese).

Wu X., Guo C., Cao Y. "A new fault diagnosis approach of power system based on Bayesian network and temporal order information". Proceedings-Chinese Society of Electrical Engineering, 2005, 25(13): 14.

Zhang C., Liao J.X., Zhu XM. "Heuristic fault localization algorithm based on Bayesian Suspected Degree". Journal of Software, 2010, 21(10): 2610–2621.

5 CONCLUSION

Using probability-weighted bipartite graph as model, Bayesian Suspected Degree as an index, the

Electrical, Control Engineering and Computer Science – Liu (Ed.)
© 2016 Taylor & Francis Group, London, ISBN 978-1-138-02937-8

Combining region-based model with geodesic active contour for nature image segmentation using graph cut optimization

Lin Song, Mantun Gao & Sanmin Wang
School of Mechanical Engineering, Northwestern Polytechnical University, Xi'an, Shaanxi, China

ABSTRACT: As the active contour model segment images using level set formulation, such formulation result in very slow algorithms that get easily stuck in local solutions and only segment image with intensity homogeneity. In this paper, a new model combining region-based with geodesic active contours is proposed for image segmentation. The new energy functional can be iteratively minimized by graph cut algorithms with high computational efficiency compared with the level set framework. Experiment results show that the proposed model can effectively and efficiently segment images with intensity inhomogeneity. The method is less sensitive to the location of initial contour and it can also avoid local minima solutions.

Keywords: image segmentation; geodesic active contours; level set method; graph cut

1 INTRODUCTION

Image segmentation is a frequent pre-processing step whose goal is to simplify the representation of an image into meaningful and spatially coherent regions (also known as segments or superpixels) with similar attributes such as consistent parts of objects or of the background (Mignotte, 2012). It plays a role in scene reconstruction, motion tracking, content-based image retrieval, aerial imaging, and medical image analysis. A large variety of segmentation algorithms have been proposed over the past few decades.

Malladi (1995) and Caselles (1997) proposed their own geometric active contour model; it is possible to automatically handle topological changes like the merging and splitting of the evolving curve. It also has the advantage that the calculation results are of high precision and robust. A review of literature indicates that most of image segmentation algorithms can be generally divided into two categories, namely the so-called region-based and edge-based active contour segmentation approaches. The edge-based models utilize image gradient to stop the curve propagation, the model can only detect target edges defined by gradient. In practice, this stopping force (edge strength) could not prevent the boundaries from leaking. Hence the propagating curves may pass through the boundary. The result of image segmentation by edge-based models is highly dependent on the initial contour placement.

In contrast to edge-based schemes, the region-based models utilize the statistical information inside and outside the contour to control the evolution, which are less sensitive to noise and have better performance for images with weak edges or without edges. The algorithm of Chan and Vese (2001) is one of the most widely used region-based methods for two-phase image segmentation. The CV model generally works for images with intensity homogeneity since it assumes that the intensities in each region always maintain constant. Thus, it often leads to poor segmentation results for images with intensity inhomogeneity due to wrong movement of evolving curves guided by global image information.

In fact, intensity inhomogeneity often occurs in real images. To solve this problem, Li et al. proposed a Local Binary Fitting (LBF) model and Region-Scalable Fitting (RSF) model (Li 2007, 2008), and the improved active contour model combining local and global information (Zhang 2010, Wang 2010, Liu 2012). To some extent, these methods are still depending on the location of initial contour and sensitive to noise.

These models mentioned above are based on partial differential equations numerical method for solving level set active contour models. They have the following two shortcomings: (1) they depend on the gradient descent method for optimization of the continuous problem, so it is easy to get stuck in local optima; (2) the effect of image segmentation depends on the number of iterations.

The graph cut method is well known in the field of computer vision. Graph cut optimization (Boykov, 2001) as a method optimization tools, it uses appropriate energy function framework to find a minimum cut in a graph. The application to the two-phase Mumford-Shah segmentation model was proposed by El-Zehiry (2007, 2011). It was shown that the graph cut is much efficient and robust than the gradient descent method.

In this paper, we present a new active contour model that utilizes the advantages of the region-based and edge-based active contour (RBCV). We use the graph cut optimization (El-Zehiry, 2011) to improve the speed and the accuracy of the minimization of the new active contour model. The use of graph cut optimization to improve efficiency has been explored by Tao, 2011. The graph cut can solve for global minimum rather than a local one and make it much less sensitive to initialization. It does not need more parameters to segment the nature image with intensity inhomogeneity.

The rest of this paper is organized as follows. In Section 2, we review well-known region-based model and graph cut. The RBCV model and the implementation are proposed in Section 3. The results are given in Section 4. This paper is summarized in Section 5.

2 RELATE WORK

2.1 The C-V model

For a given image $u(x)$ in domain Ω, the C-V model is formulated by minimizing the following energy functional:

$$E^{CV}(c_1, c_2, C) = \lambda_1 \int_{in(C)} |u(x) - c_1|^2 \, dx$$

$$+ \lambda_2 \int_{out(C)} |u(x) - c_2|^2 \, dx + \mu |C| \quad (1)$$

where c_1 and c_2 are two constants which are the average intensities inside and outside the contour, respectively.

Using the level set to represent C, that is, C is the zero level set of a Lipschitz function $\phi(x)$, we can replace the unknown variable C by the unknown variable $\phi(x)$, and the energy function can be written as

$$E^{CV}(c_1, c_2, \phi) = \lambda_1 \int_{\Omega} |u(x) - c_1|^2 H(\phi(x)) \, dx$$

$$+ \lambda_2 \int_{\Omega} |u(x) - c_2|^2 (1 - H(\phi(x))) \, dx$$

$$+ \mu \int_{\Omega} \delta(\phi(x)) |\nabla \phi(x)| \, dx$$

$$(2)$$

With the level set method, we assume

$$\begin{cases} C = \{x \in \Omega : \phi(x) = 0\}, \\ in(C) = \{x \in \Omega : \phi(x) > 0\}, \\ out(C) = \{x \in \Omega : \phi(x) < 0\}. \end{cases} \quad (3)$$

where $H(\varphi)$ is the Heaviside step function. $\delta(\phi)$ is the Dirac delta function, and $\lambda_1, \lambda_2, \mu$ are fixed parameters such that $\lambda_1, \lambda_2 > 0, \mu > 0$.

Keeping c_1 and c_2 fixed, we minimize ECV (c_1, c_2, ϕ) with respect to $\phi(x)$, and deduce the associated Euler–Lagrange equation for $\phi(x)$:

$$\frac{\partial \phi}{\partial t} = \delta(\phi) \left[\mu \nabla \left(\frac{\nabla \phi}{|\nabla \phi|} \right) - \lambda_1 (u - c_1)^2 + \lambda_2 (u - c_2)^2 \right]$$

$$(4)$$

The constants c_1 and c_2 can be reformulated using the level set function φ as:

$$c_1(\phi) = \frac{\int_{\Omega} u(x) H(\phi) \, dx}{\int_{\Omega} H(\phi) \, dx} \quad (5)$$

$$c_2(\phi) = \frac{\int_{\Omega} u(x)(1 - H(\phi)) \, dx}{\int_{\Omega} (1 - H(\phi)) \, dx} \quad (6)$$

For each level set function ϕ, the parameters c_1 and c_2 are constants that represent respectively the average intensities inside and outside the curve. It only contains global information rather than local information. If the intensities with inside C or outside C are not homogeneous, the constants c_1 and c_2 will not be accurate.

2.2 Graph cut

A cut of a graph is a partition of the vertices in the graph into two disjoint subsets utilized the Max-Flow/Min-Cut Algorithms. Constructing a graph with an image, we can solve the segmentation problem using techniques for graphs in graph theory.

An undirected graph $G = (V, L)$ is defined as a set of nodes (vertices V) and a set of undirected edges (L) that connect these nodes. As usual, the nodes is pixels, $v \in V$. There are also two specially terminals nodes s (source) and t (sink) in the graph, so, $V = \{v_1, v_2, \dots v_n, s, t\}$. $L = \{e\}$ is the set of every edge. For this kind of graph, an edge that is coming out from source s or going into sink t is called a t-link. The edges connecting the other nodes except s and t are called n-links. In Figure 1 (a), the graph is 3×3 two dimension network, w_{ij} is the weight of edge connecting node v_i and v_j. In Figure 1 (b), A cut on G is a partition of the vertices V into two disjoint subsets S and T such that $s \in S$ and $t \in T$.

108

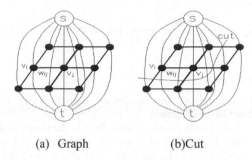

(a) Graph   (b)Cut

Figure 1. 3×3 Two dimensions graph cut.

3 THE PROPOSED METHOD

In this section, we present and discuss in detail the proposed combining region-based and edge-based model. The energy function is minimized by graph cut algorithms.

3.1 *Improved active contour*

In practice, the CV model generally fails to segment images with intensity inhomogeneity. During the level set evolution, the segmentation is easily affected by the noise. The results of segmentation cannot satisfy the need. According to LBF (Li 2007) model and LRCV (Liu 2012) model, the CV model using the two constants c_1 and c_2, replace with the two functions $c_1(x)$ and $c_2(x)$ to describe the intensities inside and outside the curve. In order to have better boundary of object, the length of the geodesic active contour models add to the last one (Tao, 2011). In the proposed model combining region-based and edge-based (RBCV), the energy function utilize the local and global information.

$$E^{RBCV}(c_1,c_2,C) = \lambda_1 \int_{in(C)} |u(x) - c_1(x)|^2 dx$$
$$+ \lambda_2 \int_{out(C)} |u(x) - c_2(x)|^2 dx$$
$$+ \mu \oint g(C) ds \qquad (7)$$

$$c_1(x) = \frac{\int_\Omega K_\sigma(x-y) u(y) H(\phi(y)) dy}{\int_\Omega K_\sigma(x-y) H(\phi(y)) dy}; \qquad (8)$$

$$c_2(x) = \frac{\int_\Omega K_\sigma(x-y) u(y)(1 - H(\phi(y))) dy}{\int_\Omega K_\sigma(x-y)(1 - H(\phi(y))) dy}; \qquad (9)$$

where y is the neighbourhood of x. For a point $x \bullet R_2$, its intensity can be approximated by a weight average of the image intensity $u(y)$. In the progress of algorithm iteration, $K_\sigma(x)$ is a Gaussian kernel function.

$$K_\sigma(x) = \frac{1}{(2\pi)^{n/2} \sigma^n} e^{-|x|^2/2\sigma^2} \qquad (10)$$

In order to decrease the effect of noise, we can use a Gaussian filtering in algorithm iterations.

The third term of Eq. (7) is the geodesic active contour $\mu \oint g(C) ds$ where

$$g = 1 \Big/ (1 + \beta |\nabla u|)^\circ \qquad (11)$$

3.2 *Graph cut optimization*

As is the case for many variational image processing models, the energy functional to be minimized is non-convex and therefore has local minima. The Chan-Vese method, as well as many other solution techniques, is based on gradient descent and is prone to getting stuck in such local minima (Ethan, 2012). In order to obtain the global optima in a relatively very short time, Eq. (7) is a typical energy functional that can be optimized by the graph cut method mentioned in (El-Zehiry, 2011).

Define a binary variable x_p for each pixel $p \bullet \Omega$ such that;

$$x_p = \begin{cases} 1 & \varnothing(p) > 0 \\ 0 & \varnothing(p) \le 0 \end{cases} \qquad (12)$$

For each point $r \bullet \Omega$, r is the neighbourhood of p, c_1 and c_2 can be represented in discrete from:

$$c_1(x) = \frac{\sum_p u(r) K_\sigma(x-r) x_r}{\sum_p K_\sigma(x-r) x_r} \qquad (13)$$

$$c_2(x) = \frac{\sum_p u(r) K_\sigma(x-r)(1-x_r)}{\sum_p K_\sigma(x-r)(1-x_r)} \qquad (14)$$

Boykov and Kolmogrov (2003) introduced the concept of cut metric in graphs that consider a cut on a grid-graph G as a closed contour. They assign the weights to the edges of the graph such that the cost of the cut approximates the length of the contour. By Cauchy–Crofton formula (Kolmogrov, 2005), the following discrete formula can be used to approximate the length of a contour,

$$\|C\|_E = \frac{1}{2} \sum_k n_k \frac{\delta^2 \Delta\theta_k}{|e_k|} = \sum_k n_k \omega_k \qquad (15)$$

where δ is the grid size, n_k is the total number of intersections of curve C with the edge lines, e_k is the shortest distance between the two nodes, ω_k is the weight of the edge e_k,

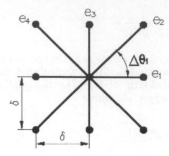

Figure 2. Neighborhood system of size 8.

$$\omega_k = \frac{\delta^2 \Delta \theta_k}{2|e_k|} \quad (16)$$

In this paper, N_8 neighbours are used to approximate the length, in Figure 2, $|e_1| = |e_3| = 1$, $|e_2| = |e_4| = \sqrt{2}$, $\varepsilon_k = \{e_1, e_2, e_3, e_4\}$, $\Delta\theta_k$ is the angular differences between the nearest families of the edges.

The discrete form of the length of the curve $\oint ds$ can be expressed as follows

$$\|C\|_E = \sum_{e_{pq} \in \varepsilon_k} \omega_{pq} \left(x_p (1 - x_q) + x_q (1 - x_p) \right) \quad (17)$$

Combing these together, we get the following discrete form of the energy function:

$$E^{RBCV}\left(c_1, c_2, \varnothing, x_1, ..., x_n\right)$$

$$= \mu \sum_{e_{pq} \in \varepsilon_k} \frac{\omega_{pq} \left(x_p (1 - x_q) + x_q (1 - x_p) \right)}{\left(1 + \beta |u(p) - u(q)| \right)}$$

$$+ \lambda_1 \sum_p |u(p) - c_1(p)|^2 x_p$$

$$+ \lambda_2 \sum_p |u(p) - c_2(p)|^2 \left(1 - x_p\right) \quad (18)$$

In order to use graph cut algorithms, we need to construct a proper graph $G = (V, L)$. Each pixel in the image domain is considered as one node of G. Each pixel node p has two t-links, i.e. (s, p) and (p, t) respectively connecting it to the source and sink nodes and the weights are respectively defined as w_{sp} and w_{pt}. Each pair neighbouring pixels $\{p, q\}$ in the neighbour system is connected by an n-link and the weight is defined as w_{pq}. The weights of G are set according to the following equations:

$$w_{sp} = \lambda_1 |u(p) - c_1(p)|^2, \ p \in \Omega, \quad (19)$$

$$w_{pt} = \lambda_2 |u(p) - c_2(p)|^2, \ p \in \Omega, \quad (20)$$

$$w_{pq} = \frac{\mu \omega_{pq}}{\left(1 + \beta |u(p) - u(q)|\right)}, \ p \in \Omega, \forall q \in N_8(p). \quad (21)$$

3.3 Algorithm

To summarize, the integrated algorithm can be described as follows:
1. Initialize the curve C anywhere in the image and get the interior of C and the exterior of C. For each pixel $p \in \Omega$, the binary variable x_p is initialized by assigning 0 if p is outside C and 1 if p is inside C.
2. Calculate $c_1(p)$ and $c_2(p)$ using Eqs. (13) and (14).
3. Construct graph G by Eq. (19)–(21).
   For each pixel $p \in \Omega$, add an edge Sv_p with weight w_{vp} and add an edge $v_p T$ with weight w_{pt}.
   For each pixel q in the 8-neighborhood system of p calculate w_{pq} using Eq. (21), and add an edge $v_p v_q$ with weight w_{pq}
4. Compute the minimum cut of G using graph cuts algorithm and get the binary label $\{x_p | x_p \in \{0,1\}, p = 1, ..., N\}$,
5. Repeat steps 2–4 until the energy is minimized.

4 EXPERIMENTAL RESULTS

In order to verify the feasibility and effectiveness of this proposed model, a large number of natural image segmentation experiments. All the experiments were conducted in Matlab R2013 and Microsoft Visual studio 2010 programming environment on 2.80HZ Intel Pentium CPU and 2GB memory. We set the parameters $\lambda_1 = \lambda_2 = 1$, $\mu = 1500$, $\beta = 0.03$.

4.1 Less sensitive to the location of initial contour

In Figure 3, the first row is the different location of initial contour and the second row is final contour for corresponding image.

In Figure 4, the first row is the different size of initial contour and the second row is final contour for corresponding image. According to Figure 3 and Figure 4, we know that the segmentation results of image are almost same, changing the location and size of initial contour.

In Figure 5 we plot the value of the energy functional versus iteration number for the images tested in Figure 3 and Figure 4. Each plot has three curves which correspond respectively to different location and size of the initial contour. From the curves we can see that the energy function values converge to lower after two iterations of the RBCV model, after five iterations of the changes is rela-

110

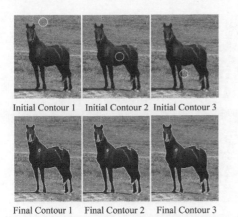

Initial Contour 1   Initial Contour 2   Initial Contour 3

Final Contour 1   Final Contour 2   Final Contour 3

Figure 3. Results of our method for the different location of initial contour.

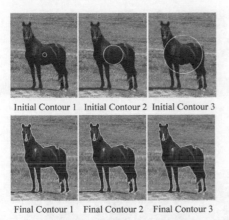

Initial Contour 1   Initial Contour 2   Initial Contour 3

Final Contour 1   Final Contour 2   Final Contour 3

Figure 4. Results of our method for the different size of initial contour.

tively small, substantially after ten iterations can achieve the minimum energy function value.

As discussed above, the location and the size of the initial curve has little effect on the segmentation results. The proposed method is less sensitive to the location and size of initial contour.

4.2 Shorten the runtime

Table 1 lists the number of iterations and CPU time of RBCV model and C-V model which segment the image shown in Figure 6 For both of the initial contours, the C-V model has less average time, but has more iterations and CPU time than the proposed RBCV model.

4.3 Effective of result

Figure 6 shows the segmentation results of nature image by two different methods. The first row

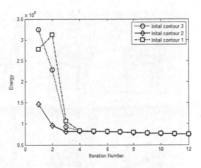

(a) Different location

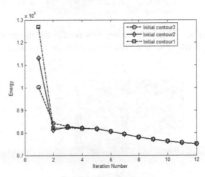

(b) Different size

Figure 5. Energy values of the different initialization.

Table 1. Comparison of the number of iterations and CPU time.

	RBCV	CV
Iterations	12	5000
Total CPU time (s)	13.102	713.065
Average time (s)	0.916	0.143

shows the iterative process by RBCV model, followed by the initial curve position, 1, 2, 7, 10, 12 times iterations of the segmentation image. The second row shows the iterative process by CV model, followed by the initial curve position, 10, 20, 500, 2000, 5000 times iterations of the segmentation image. From the result of CV model shown in Figure 6, the shadow of grass that belongs to background is labelled as the foreground. To the contrary, the reflecting portion on the horse that belongs to foreground is labelled as the background. The right segmentation result cannot be obtained from the CV model. With the same initial, using RBCV model, the boundary of the object is relatively clear and accurate.

In order to obtain an evaluation of the segmentation results, we compared the performance of

Figure 6. The comparisons of results for image by RBCV model and CV model. Row 1: The curve evolution process using RBCV. Row 2: The curve evolution process using CV.

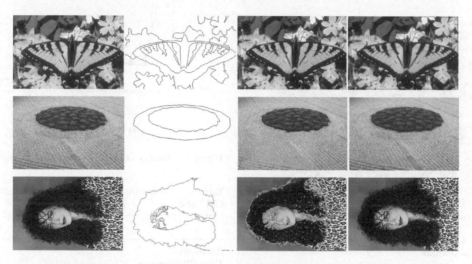

Figure 7. Results for nature image by RBCV model and CV model. Column 1: original images. Column 2: Ground truth. Column 3: final segmentation results using CV model. Column 4: final segmentation results using RBCV model.

algorithm on the nature image of Berkeley Segmentation Data Set (BSDS). In Figure 7, we compare the results for nature image by RBCV model and CV model. The original images and ground truth (BSDS) are shown in columns 1 and 2, the final segmentation results using CV model and RBCV model are shown in columns 3 and 4.

From the results we can see that, the CV model cannot segment the nature image with intensity inhomogeneity. It is sensitive to noise and easy to get stuck in local minimum. Therefore, accurate result is obtained hardly. RBCV model is insensitive to noise in the process of segmenting nature image and avoid getting stuck in local minimum. The object contours, with smooth and accurate, are obtained.

5 CONCLUSION

In this paper, we proposed an improved active contour model for image segmentation. The new model can be optimized by graph cut algorithms and can segment the nature image with intensity inhomogeneity. As the global minimization of graph cut algorithms utilized, the speed and the accuracy of the implementation are greatly improved, and make the result much less sensitive to the location of initial contour. The model has the defect that cannot simultaneously detect multiple objects in different intensities. Because nature images often contain multiple objects, in our future work, we will extend the current model by using multiple level set functions.

REFERENCES

Boykov Yuri Y., Marie-Pierre Jolly. Interactive Graph Cuts for Optimal Boundary & Region Segmentation of Objects in N-D Images. IEEE Conference on Computer Vision, 2001, 105–112.

Boykov Y., Kolmogorov V. Computing geodesics and minimal surfaces via graph cuts. IEEE International Conference on Computer Vision, 2003.

Caselles V, Kimmel R, and Sapiro G. Geodesic active contours. International Journal of Computer Vision, 22(1997): 61–79.

Chan T.F and Vese L. Active contours without edges. IEEE Transactions on Image Processing, 10(2001): 266–77.

El-Zehiry N., Xu S., Sahoo P., A. Elmaghraby, Graph cut optimization for the Mumford–Shah model, in: The Seventh IASTED International Conference on Visualization, Imaging and Image Processing, 182–187, ACTA Press, 2007.

El-Zehiry N., Sahoo P., A. Elmaghraby, Combinatorial Optimization of the piecewise constant Mumford-Shah functional with application to scalar/vector valued and volumetric image segmentation. Image and Vision Computing, 2011, 29(6): 365–381.

Ethan S., Brown Tony F., Chan Xavier Bresson, Completely Convex Formulation of the Chan-Vese Image Segmentation Model, International Journal of Computer Vision, 98(2012): 103–21.

Kolmogorov V., Boykov Y., What metrics can be approximated by geo-cuts, or global optimization of length/area and flux [J]. Proceedings of the Tenth IEEE International Conference on Computer Vision, vol. 1, 2005, 564–571.

Li Chunming, Kao Chiu-Yen, John C. Gore, Zhaohua Ding. Implicit active contours driven by local binary fitting energy. IEEE Conference on Computer Vision and Pattern Recognition, 2007, 1–7.

Li Chunming, Kao Chiu-Yen, Gore J, Zhaohua Ding. Minimization of Region-scalable Fitting Energy for Image Segmentation. IEEE Transactions on Image Processing, 17(2008): 1940–49.

Liu Shigang, Peng Yali A local region-based Chan-Vese model for image segmentation. Pattern Recognition, 45(2012): 2769–99.

Mignotte Max. MDS-based segmentation model for the fusion of contour and texture cues in natural images, Computer Vision and Image Understanding. 116(2012): 981–90.

Malladi R, Sethian J.A, Venmuri B.C. Shape modeling with front propagation: a level set approach. IEEE Transactions on Pattern Analysis and Machine Intelligence, 17(1995): 158–74.

Tao Wenbing, Tai XueCheng, Multiple piecewise constant with geodesic active contours (MPC-GAC) framework for interactive image segmentation using graph cut optimization, Image and Vision Computing, 29 (2011): 499–508.

Wang Xiao-feng, Huang De-shuang, Xu Huan. An Efficient Local Chan-Vese Model for Image Segmentation. Pattern Recognition, 43(2010): 603–18.

Zhang Kaihua, Zhang Lei, Song Huihui, et al. Active contours with selective local or global segmentation: a new formulation and level set method. Image and Vision Computing, 28(2010): 668–76.

Computer science

Electrical, Control Engineering and Computer Science – Liu (Ed.)
© 2016 Taylor & Francis Group, London, ISBN 978-1-138-02937-8

Research and implementation of automobile ECU bootloader self-update

Ji Zhang, Xiangyu Zhu & Yong Peng
Clean Energy Automotive Engineering Center of Tongji University, Shanghai, China

ABSTRACT: With the development of the automotive electron, automotive ECU remote update will be a tendency in the near future. This paper researched and realized automotive ECU Bootloader design based on the development tendency of automotive ECU remote update. By analyzing the present technology of Bootloader, a new Bootloader self-update scheme is proposed. Combined with the Bootloader technology, the new scheme is elaborately designed. Finally, the experiment result shows that the new method is safer and more efficient than the traditional method.

Keywords: ECU; remote update; bootloader; self-update

1 INTRODUCTION

With the development of automobile electronic technology, ECUs (Electronic Control Unit) and software codes in automobiles are increasing sharply. Automobile ECU firmware update becomes more and more difficult. The traditional update method cannot satisfy the challenges in the future. To implement automobile ECU firmware remote update, it is generally divided into two steps. [1] [2] In the first step, software update package is downloaded from sever to vehicle platform through the wireless network. In the second step, software update package will be programmed to target ECU though the vehicle's network. In this step, ECU should have the ability to achieve online update. To design automobile ECU Bootloader, which has the ability to achieve online update, it is necessary to implement the automobile ECU firmware remote update and it is also the key point of this paper. [3]

Some companies are doing research on automotive ECU remote update. Symphony Teleca cooperated with Movimentog Company providing the InSight Connect VRM technology program and Venturo technology program in 2013, which is a safe and effective solution for ECU remote update [4], Red Bend Company also cooperated with Vector to use wireless downloader technology (FOTA) for supporting automotive remote update.

According to the current automobile networking communication technology, the ECU remote update system selects 3G mobile network technology as the way of wireless communication and the automobile platform chooses the scheme which separates wireless download function and Human Machine Interface (HMI) from ECU update function. Update software is downloaded from Internet to smart mobile phone terminal. The HMI is also designed in mobile phone which can receive update software from Internet and sent software to the embedded vehicle platform by wireless net. During the process of Software update, some criterions, such as OSEK, AUTOSAR, could be referred to ensure that software update is safe and reliable.

2 ANALYSIS OF AUTOMOBILE ECU BOOTLOADER TECHNOLOGY

2.1 *Bootloader summary*

Bootloader is also called boot load program. This program codes is executed firstly after turn on the hardware equipment. For different application fields, the work methods are also different.

In general, vehicle ECU does not need to load the operating system kernel, the main job of the Bootloader is used to update application program codes, namely the online update function.

Online update function is an implementation of the IAP (In Application Programming) technology. When the program is running, program memory can be erased by the program itself. Simply put, the procedures could write date or modify data into program memory by itself. This procedure, which is programming, is called the application program. The implementation of this section programming function is Bootloader program. Now the Flash storage technology is widely used in the automobile ECU to erase or write flash storage technology support software, provided the automobile ECU has the ability to achieve online updates.

In general, the entire realization function of Bootloader is dependent on the hardware platform, especially in the embedded system. So it is impossible to design a universal Bootloader code. Different CPU structures have different Bootloader programs. Even if a Bootloader program could run in a circuit board, the Bootloader source codes should be amended if you want to run it in another board.

2.2 Memory layout

In order to realize the online-update function, ECU memory space should be divided into two parts first. One part is the Bootloader area, and another is application program area. Two-part codes are completely independent. The application program area includes the interrupt vector remapping table and specific application code. In order to ensure that the memory works efficiently, unnecessary memory gaps should be eliminated. Generally, the Bootloader area is adjacent to the interrupt vector table.

3 RESEARCH OF BOOTLOADER SELF-UPDATE SCHEME

Bootloader self-update means that through its own ability of online update, it can communicate with the host computer and download the updater, to complete self-update job.

3.1 Basic solution scheme

The feature of Bootloader self-update solution scheme layout is that the Bootloader area is connected closely with the interrupt vector table. This purpose could reduce memory fragmentation and improve memory utilization.

As shown in Figure 1, the basic process of the Bootloader self-update is described as follows.

(Updater contains downloaded components and flash driver; they could program and erase the designated Bootloader area):

1. After system reset, "startup manager" judges that Bootloader update-request enters the Bootloader update service.
2. The designated area is erased first in the process of Bootloader update service, and the updater is downloaded from the host computer, and the designated area is then programmed.
3. Jumping to the execution area of the updater, the updater downloads the latest version of Bootloader from the host computer, and completes the corresponding update job. During the execution process of the above basic scheme, external abnormal situations (power down) happen which will make the whole system go into a "isolation" state. As shown in Figure 2, the red time zone belongs to "danger zone." In this zone, if abnormal situations happen before re-programming the reset vector (because Bootloader area is close to interrupt vector table, during the process of Bootloader re-programming, reset vector will also be erased and re-programmed), system will lose the reset vector. If abnormal situations happen before it, due to incomplete writing of the Bootloader area, it will also cause the phenomena of program fleet.

3.2 Improved scheme

In order to solve the above problems, two things need to be done:

1. Bootloader area is isolated with interrupt vector table and stored in different blocks, which could be erased;
2. Add "boot exchange sign" and do the reset vector re-mapping by this sign.

Improved memory space layout is shown in Figure 3; the "boot exchange sign" must be stored in a nonvolatile memory space. The process of resetting the interrupt vector re-mapping table is very simple. When the "boot exchange sign" is valid, jump to the "updater starting address."

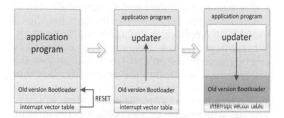

Figure 1.   The basic scheme of Bootloader self-update.

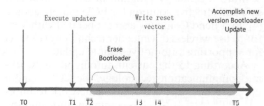

Figure 2.   The time line of Bootloader self-update.

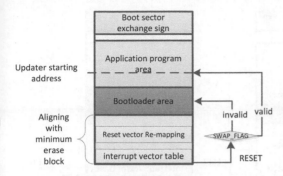

Figure 3. Improvement of ECU memory space layout according to Bootloader self-update.

Otherwise, enter the Bootloader area. The remapping table is different from the application program interrupt vector table. It must remain unchanged in the update process.

The self-update process of the Bootloader, which is based on this memory layout, is shown in Figure 4. The basic process is described as follows:

1. After the first time reset, jump to reset interrupt vector re-mapping area. By set "boot exchange sign" to invalid, jump to the Bootloader area and start processing Bootloader update request that enters Bootloader update service. In the process of update service, the designated application area is erased first, then the updater is downloaded from the host computer and programmed in the designated area. Finally, set the "boot exchange sign" to valid. The software enables system reset.
2. In the second time reset of system, reset the interrupt vector remapping table by setting "boot exchange area" to valid. Jumping to the updater, the updater will communicate with the host computer and accomplish the new version update of Bootloader; the "boot exchange sign" will also be set to invalid.
3. In the third time reset of system, reset interrupt vector remapping table by setting "boot exchange sign," to invalid. Jump to the new version of Bootloader area. Start processing the update request to enter the Bootloader update service. Complete the re-programming of application program during the update service. Because Bootloader is isolated with interrupt vector in the scheme, the reset vector is also isolated. Therefore, there is no need to re-write the reset vector in the whole process. At the same time, through the reset vector re-mapping, it can ensure that re-mapping will enter the effective execution area even if abnormal situations happen at any time during the update process,

thereby avoiding the whole system entering into "isolated" state.

3.3 A new solution scheme

Based on the above scheme, this paper proposes a new solution. The memory space layout of new scheme is shown as Figure 5. Compared with the previous schemes, this scheme divides Bootloader into two blocks. Bootloader self-update process based on this memory space layout is shown as Figure 6. The basic process is described as follows:

1. After system reset, jump to reset interrupt vector remapping table, setting "boot exchange sign" to invalid, and jump to the Bootloader area 1. Start processing Bootloader update request to enter the Bootloader update service. In the update service, erasing the Bootloader area 2 first. Then the new version of Bootloader will be downloaded from the host computer programmed in the designated location. Finally, set the "boot exchange sign" to valid, and the software then enables the system to reset.
2. When system reset for the second time, programming jump to reset interrupt vector remapping table. Setting "boot exchange sign" to invalid, jump to the Bootloader area 2, i.e., a new version of the Bootloader area.

The new scheme is easy to implement. By isolating Bootloader and reset interrupt vectors, to prevent losing the reset vector, use reset vector remapping to switch Bootloader area 1 and area 2. When abnormal situations occur at any time during the update process, it can ensure the remapping to the intact Bootloader area, preventing the whole system from entering the "isolated" state.

3.4 The final state of ecu layout

The final ECU memory layout is shown in Figure 7; it could effectively support the Bootloader self-update function and prevent the system from entering into the isolated state. In the layout, there are: application program update area, Bootloader area, and a fixed area which is located in the program/ Flash code. These three block areas should keep the program/code aligned with minimum flash erased block. The layout assigns a data flash block or EEPROM area, which is used as the user data storage, wherein the backup area is effective when ECU doesn't contain the EEPROM. Due to lack the support of EEPROM, it cannot perform single byte operations by the date; when rewriting data, the entire sector need to be erased. If abnormal situations happen; it will lose data, so it needs to distribute to another area as an information backup.

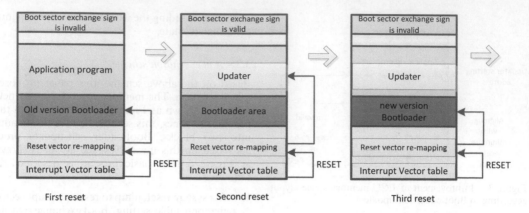

First reset   Second reset   Third reset

Figure 4. Improvement scheme process of Bootloader self-update.

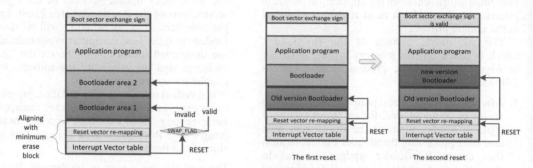

Figure 5. A new ECU memory space layout of Boot-loader self-update scheme.

Figure 6. A new scheme process of Bootloader self-update.

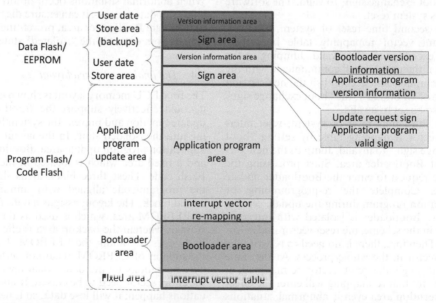

Figure 7. Memory space layout.

The user data store area includes two parts: the version information area and sign area. The version information is used to store the Bootloader version information and application version information. In the version validation, this part of information need to be read and compared with the version information, which is sent from the host computer; while the update is completed, version information will be re-written. In order to prevent external abnormity, this part of information needs to be stored separately to retain its effectiveness, e.g., if the system loses power suddenly during the process of Bootloader update, after erasure operation is executed. If the version information is stored in the Bootloader area, the part of the information will be lost. When system power comes on and update job is executed again, it will not read the version information and version validation job will not be performed. The update request sign in the sign area is used to enter the Bootloader update service from the application by a system reset. The effective sign of application program is used to prevent the abnormal situations during the process of application program update; the boot area exchange sign is used to ensure reliable update of Bootloader.

3.5 State flow

The state of flow of ECU nodes is shown in Figure 8, involve five running states: application program running state, system initialization state, Bootloader alarm state, Bootloader locked, and unlocked state, out of which four states are running in the Bootloader area.

3.6 The specific operation process

The specific operation process of ECU update service, as shown in Figure 9, executes Bootloader

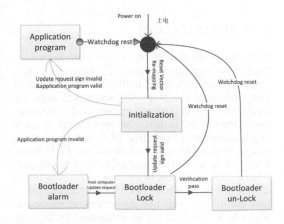

Figure 8. State flow.

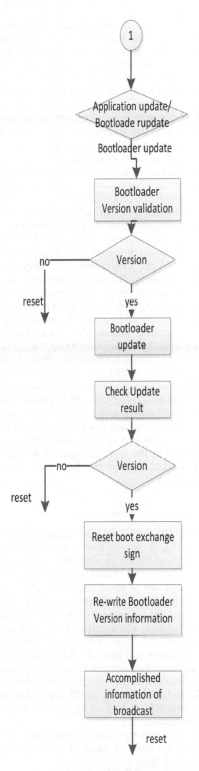

Figure 9. Specific operational process of ECU update service.

update. At first, communication with the host computer executes Bootloader version validation job. If the validation does not pass (update software package version is not newer than the current version), it will enable the watchdog reset. If the validation passes, the entire Bootloader update job will be executed. After the update job is completed, check whether the update result is correct (update results are achieved by calculating CRC). If the validation result does not pass, then enable watchdog reset. If it passes, the reset boot sign will switch (after system reset, switch to Bootloader area which has been updated according to this sign) and re-write Bootloader version information in the user data area. Finally, broadcast update complete message to the other on the ECU network.

The specific operation process of ECU update service, as shown in Figure 9, executes Bootloader update. At first, communication with the host computer executes Bootloader version validation job. If the validation does not pass (update software package version is not newer than the current version), it will enable the watchdog reset. If the validation passes, the entire Bootloader update job will be executed. After the update job is completed, check whether the update result is correct (update results are achieved by calculating CRC), if the validation result does not pass, then enable watchdog reset. If it passes, the reset boot sign will switch (after system reset, switch to Bootloader area which has been updated according to this sign) and re-write Bootloader version information in the user data area. Finally, broadcast updated complete message to the other on the ECU network.

3.7 Proving of scheme

In order to verify the Bootloader self-update scheme, this paper produces two different versions of Bootloader in the BT1 area and BT2 area: Bootloader1.s19 and Bootloader2.s19. The Bootloader1.s19 is based on the RS232 communication; Bootloader2.s19 is based on CAN communication. So it can verify whether the function is correct or not. The specific flow is based on the "CAN communication of the host computer software," as shown in Figure 10. First, turn on USB-CAN device, then set the related parameters of the CAN communication, including the selected channel and the target ECU ID, then select the target file which is desired to be update, and finally start the update.

After the software update package update is successful, the running result of the host computer is shown in Figure 11.

Tests prove that Bootloader online update system based on the CAN bus could effectively

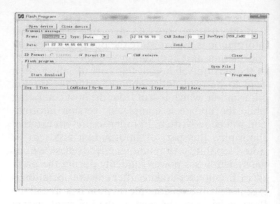

Figure 10.   The using process of the host computer software "Flash Program".

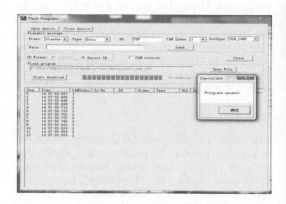

Figure 11.   After software update is successful, the running effect of host computer software.

complete application program of Bootloader self-update.

4   SUMMARIZATION

With the development of automotive electronics, the software update of auto ECU becomes a challenge. Current updating methods could not meet this challenge; therefore, the implementation of automotive ECU remote update is the development trend in the future. This paper aims to study applications of automotive ECU remote update, design, and implement the automotive ECU Bootloader. A detailed study and analysis of Bootloader related technology has been done, according to the design requirements of ECU Bootloader, and a new Bootloader self-update scheme is proposed. The scheme implements an integrated ECU automobile Bootloader update system. After going through the test, the new scheme is proved to be safe and reliable.

REFERENCES

[1] Miucic R., Mahmud S.M. Wireless multicasting for remote software upload in vehicles with realistic vehicle movement [C]. Proc. SAE World Congress, 2005: 11–14.

[2] Lightner, B., Botrego, D., Myers, C., Lowrey, L.H. Wireless diagnostic system and method for monitoring vehicles. U.S. patent 6636790, 2003.

[3] Rezgui J., Cherkaoui S., Charkroun O. Deterministic for DSRC/802.11p vehicular safety communication [C]. 2011 7th International Wireless Communications and Mobile Computing Conference, 2011, p 595–600.

[4] MOUNTAIN VIEW, Calif. Symphony Teleca and Movimento Partner to Provide Advanced Over-the-Air ECU Software Update Solution to the Automotive Market[EB/OL]. http://www.movimentogroup.com/07-november-2013/, 2013.

[5] Nilsson D.K., Sun L., Nakajima T. A framework for self-verification of firmware updates over the air in vehicle ECUs [C]. GLOBECOM Workshops, 2008 IEEE. IEEE, 2008: 1–5.

[6] Thanthry Sreedhar, S. Remya Technique for Reprogramming the Boot in Automotive Embedded Controllers [C]. SAE World Congress. 2008-01-0384.

[7] Miucic R., Mahmud S.M. Mobile Multicasting for Remote Software Update in Intelligent Vehicles [J].

[8] Miucic R., Mahmud S.M. Wireless Reprogramming of Vehicle Electronic Control Units [C]. Consumer Communications and Networking Conference, 2008. CCNC 2008.5th IEEE. IEEE, 2008: 754–755.

[9] Tan T., Tang H., Zhou Y. Design and Implementation of Bootloader for Vehicle Control Unit Based on Can Bus [C]. Proceedings of the FISITA 2012 World Automotive Congress. Springer Berlin Heidelberg, 2013: 447–457.

Electrical, Control Engineering and Computer Science – Liu (Ed.)
© 2016 Taylor & Francis Group, London, ISBN 978-1-138-02937-8

An embedded web server for remote monitoring of rice whiteness

K. Kanjanawanishkul, J. Chinnakotr & W. Promwang
*Mechatronics Research Unit, Faculty of Engineering, Mahasarakham University, Kamriang,
Kantharawichai, Mahasarakham, Thailand*

ABSTRACT: This paper focuses on a device that can measure whiteness percentage of rice grains after whitened by a rice-whitening machine. Our device has the similar performance to the commercial whiteness meter, as shown in experimental results. Thus, it will be installed at a rice mill in the future in order that an operator can monitor the whiteness percentage of rice grains during the whitening process and these data can be monitored by an owner or a manager anywhere. The overall system consists of two parts. The functions of the first part, called a sensor node, are to estimate the whiteness percentage of rice grains by using data from a color sensor and to transmit it with other information through wireless networks. In the second part, an embedded web server gateway, consisting of a Raspberry PI single board computer and an Arduino development board, receives and records all information from the sensor node. Furthermore, it delivers a web page to web clients, if there is a request from web clients. All information from the sensor node can be remotely monitored via the web browser as seen in our experimental results.

Keywords: rice whiteness; web server; color sensor; remote monitoring; sensor node

1 INTRODUCTION

Whiteness of rice grains is a key factor for making a decision to increase or decrease the power in a rice whitening process. Traditionally, several large-scale rice mills in Thailand monitor milled rice quality manually at sampling interval of 1–2 hours since there are no continuous on-line measurement methods available. In general, the whiteness of rice is measured by a commercial whiteness meter, e.g. model C-300 of Kett Electric Laboratory, where the sample is scanned with filtered light and the color value (0–100) is calculated and shown on the display. However, this meter is quite costly and it is needed to be controlled to get an accurate whiteness value. In some cases, an experienced operator/manager determines subjectively the whiteness through visual inspection or by holding a hand over rice grains after whitened to estimate the temperature of rice since there is a relationship between the whiteness and temperature of whitened rice. However, this method is improper for unexperienced operators.

In general, whiteness is the degree to which a surface is white. A commercial whiteness meter is designed with the demand of CIE, standard light resource and light environment (Lei et al. 2008). In the literature, there are a number of investigations that propose different ways to measure the whiteness of rice grains. Jangkajit & Khunboa (2013) developed a device installed at the whitening step

to record images and measure the temperature of rice flowing along the groove. They found that the temperature can be used to represent the whiteness value of rice. Kawamura et al. (2003) built an automated method that can measure whiteness values of rice by using a reflectance sensor. Nascimento & Galli (2008) built a device that measured the whiteness of rice grains by measuring the reflected light on a rice sample from light sources. Yadav & Jindal (2001) estimated whiteness of the milled rice samples using image analysis from measurements of grey level distribution of an image of milled rice grains. Liu et al. (1998) used digital image analysis to measure degree of milling expressed quantitatively as surface lipids concentration of rice grains.

Since the milling degree is correlated with the whiteness level of rice grains, the first contribution of this paper is to provide a low-cost and ease-of-use device with reasonable accuracy and repeatability of measurements using a color sensor, compared with the commercial whiteness meter. The second contribution of this paper is remote monitoring, recording and online web browsing. To achieve these, the whiteness level of rice grains is transmitted to an embedded web server via wireless communication. These data are then recorded and can be retrieved by the embedded web server as requested by a web browser. Therefore, they can be accessed anywhere and anytime.

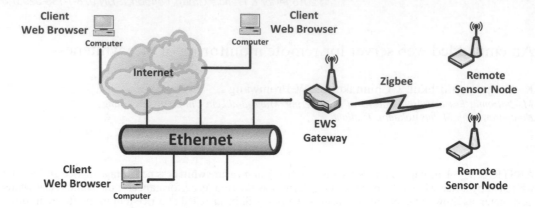

Figure 1. A remote monitoring system model.

2 PROPOSED METHODS

The contribution of this paper is twofold. The first one is to provide a device that can estimate the whiteness value of rice grains, while the second one is to implement remote monitoring through wireless networks and online web browsing showing data received from our developed device. This can be achieved by using two main parts, i.e. a sensor node and an Embedded Web Server (EWS) gateway, as shown in Figure 1. Both are connected through a Zigbee wireless network.

2.1 Sensor node

A block diagram of the sensor node is shown in Figure 2.

The main component of the sensor node is a TCS230 color sensor with 4 white LEDs. It consists of four types of photodiodes, i.e. blue, green, red, and clear. Spectral responsivity of each photodiode can be found in its datasheet. Four white LEDs shine down at 45 degrees onto the sample to flood it with white light. This is called geometry 45°/0°, i.e. the sample is illuminated in an angle of 45° and observed under 0° from the normal. Then the output of the color sensor is a square wave with the frequency directly proportional to light intensity.

In this work, to measure the whiteness value, milled rice samples are placed into a cylindrical sample holder with a diameter of 53 mm and a depth of 8 mm as shown in Figure 3b. To reduce reflection of the ambient light, the internal surface of the enclosure were painted flat black. The whiteness level is measured by capturing the reflected light, on a rice sample from LED light sources.

An Arduino Mega 2560 board is employed to process frequency data from the sensor so that

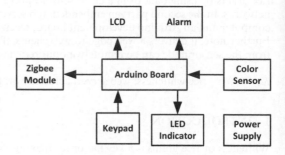

Figure 2. A block diagram of the sensor node.

the whiteness level is estimated, displayed on the LCD as seen in Figure 3a and transmitted via the Zigbee wireless network to the EWS gateway. Furthermore, the operator can specify the upper and lower allowable limits of the whiteness level. If the estimate of the whiteness level is out of the given allowable range, the red LED and the alarm indicator will be triggered to alert the operator.

2.2 Embedded Web Server (EWS) gateway

A EWS gateway is used to link two network systems, i.e. the Zigbee wireless network for the sensor node and the Ethernet network for the Internet. A block diagram with main components is shown in Figure 4. The EWS gateway receives and records all data from the sensor node and acts as a web server that can provide web pages to any web browsers from anywhere.

To realize such functions, an Arduino Mega 2560 board, called a wireless sensor interface board, is employed to collect all data from the sensor node and then transmits to a Raspberry PI single board computer via serial communication. The fundamental services required to turn the Raspberry

126

(a)

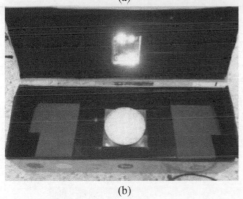

(b)

Figure 3. Our device for whiteness measurement using a TCS230 color sensor: (a) its front panel with an LCD display, and (b) its inside look.

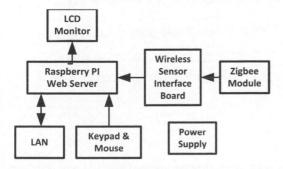

Figure 4. A block diagram of the EWS gateway.

Figure 5. A EWS gateway.

Pi (see Fig. 5) into a EWS consist of Apache (the web server itself), PHP (scripting language), and MySQL (database server).

3 EXPERIMENTAL RESULTS

To show the effectiveness of our proposed method, the following two experiments have been conducted.

3.1 Performance tests of the proposed device

The rice used for this study consisted of 11 samples with different milling degrees. The first 5 samples were used to establish the relationship between the responses of the RGB color sensor and the whiteness percentage taken from the commercial whiteness meter (model C-300, Kett Electric Laboratory), while the other samples were used to test the accuracy of the equation. All measurement steps on the whiteness level of milled rice samples were repeated three times by filling the sample holder each time to attenuate the random error. A white plate was used for white balance adjustment of the TCS230 color sensor. The experiment was conducted as follows:

1. The RGB intensity values (ranging from 0 to 255) of the first 5 rice samples measured by our device were compared with the corresponding whiteness percentage indicated by the commercial whiteness meter. The results are presented in Table 1.
2. The linear regression technique was used to determine the relationship of the results shown in Table 1. There is a strong linear relationship between the whiteness meter reading and the RGB intensity values as indicated by a high coefficient of multiple determination ($R^2 = 0.9999$). That relation exists as follows:

$$y = 2.7282 - 0.2105\,(x_1) + 0.6319\,(x_2) - 0.0596\,(x_3) \tag{1}$$

127

where y is the estimate of the whiteness percentage of rice grains and x_1, x_2, and x_3 are the intensity values of red, green, and blue, respectively.

3. The other six milled rice samples were used to verify the accuracy of our device. In Table 2,

Table 1. Whiteness percentage of milled rice samples measured by the commercial whiteness meter served as the reference values and the RGB intensity values of those determined by our device.

Rice samples	Whiteness percentage by C-300 Kett whiteness meter [%]	RGB intensity values by our device		
		R	G	B
1	35.16	105	93	71
2	37.90	108	99	78
3	40.80	116	107	86
4	43.00	120	112	88
5	44.80	125	117	93

it is found that the estimated values of milled rice whiteness based on Equation 1 in terms of whiteness percentage were close to the measurements by the commercial whiteness meter.

3.2 Remote monitoring tests via the embedded web server

In the second experiment, remote monitoring via the web browser was evaluated as follows:

1. A MySQL database was first created. It contains the ID number as a primary key, current timestamp, whiteness percentage, and upper and lower allowable limits.
2. A program of the sensor node was executed to estimate the whiteness percentage and to transmit all data to the EWS gateway via wireless communication.
3. A program of the wireless sensor interface board was executed to receive all data from the sensor node via wireless communication and to

Table 2. The other six rice samples tested on our device compared with the commercial whiteness meter.

Rice samples	Whiteness percentage by C-300 Kett whiteness meter [%]	RGB intensity values by our device			Whiteness percentage based on Equation 1 [%]	% error
		R	G	B		
1	41.76	118	108	85	41.56	0.48
2	42.40	110	107	87	42.06	0.8
3	43.90	120	110	91	43.20	1.06
4	43.96	119	112	92	43.94	0.45
5	44.10	115	106	86	43.85	0.57
6	44.50	119	112	97	44.78	0.62

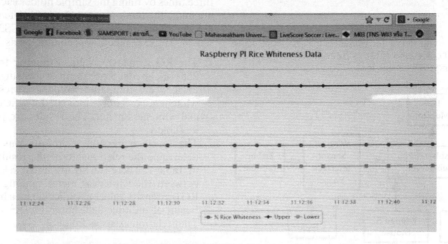

Figure 6. Three lines in the chart displayed on the web browser representing the estimated whiteness percentage, the upper, and the lower allowable limits of the whiteness percentage.

transfer all received data to the EWS via serial communication.

4. A program written in Python on the Raspberry PI was executed to receive the data from the wireless sensor interface board via serial communication and then all data were recorded into the database created at step (1).

If there is a URL request from the web browser, the EWS retrieves data from the MySQL database and then creates a web page to remotely display line charts of the estimated whiteness percentage, and the upper and the lower allowable limits on the web browser. These line charts shown in Figure 6 were implemented by using Highcharts (http://www.highcharts.com), which is a charting library written in HTML5/JavaScript.

4 CONCLUSIONS AND FUTURE WORK

Due to consumer preference of white rice grains, the degree of milling determines the amount of removal of bran layer from the surface of milled rice and thus relates to its whiteness. Because of this, our device has been built to estimate the whiteness percentage of the milled rice. As seen in the experimental results, our developed device has no sensible deviation from the measurements given by a commercial whiteness meter. Thus, this is an opportunity to use our low-cost and reliable device in mill industries. For remote monitoring, whiteness percentage of rice grains and its allowable percentage range can be transmitted via wireless communication from the sensor node to the EWS and remotely displayed on the web browser successfully.

In the future, this device can be used to continuously monitor whitened rice during the whitening process, resulting in improving the milling process and quality assessment. Development of wireless sensor networks for the rice mill is also our future research direction.

ACKNOWLEDGEMENTS

The authors would like to thank the Post-Harvest and Agricultural Machinery Engineering Research Unit, Faculty of Engineering, Mahasarakham University for allowing us to use the laboratory equipment and resources.

REFERENCES

Jangkajit, C. & Khunboa, C. 2013. The study of relationship between whiteness and temperature of rice, Proceedings of the 2nd International Conference on Advances in Computer and Information Technology (ACIT 2013), Kuala Lumpur, Malaysia, 2013.

Kawamura, S., Natsuga, M., Takekura, K. & Itoh, K. 2003. Development of an automatic rice-quality inspection system. *Computers and Electronics in Agriculture* 40(1–3): 115–126.

Lei, H., Ruan, R., Fulcher, R. & Lengerich, B. 2008. Color development in an extrusion-cooked model system. *International Journal of Agricultural and Biological Engineering* 1(2): 55–63.

Liu, W., Tao, Y., Siebenmorgen, T. & Chen, H. 1998. Digital image analysis method for rapid measurement of degree of milling of rice. *Cereal Chemistry Journal* 75(3): 380–385.

Nascimento, T. & Galli, R. 2008. An equipment to measure whiteness and transparency of rice. *Revista Ciências Exatas* 14(2): 1–7.

Yadav, B.K. & Jindal, V.K. 2001. Monitoring milling quality of rice by image analysis. *Computers and Electronics in Agriculture* 33(1): 19–33.

Electrical, Control Engineering and Computer Science – Liu (Ed.)
© 2016 Taylor & Francis Group, London, ISBN 978-1-138-02937-8

Detection of gene deletion based on machine learning

Yan Zhao, Jingyang Gao, Fei Qi & Rui Guan
*College of Information Science and Technology, Beijing University of Chemical Technology, BUCT,
Beijing, P.R. China*

ABSTRACT: In bioinformatics, the detection of gene deletion has been an important research direction. In this paper, machine learning is combined with biology, and is applied to detect gene deletion. Based on the human chromosome 21 and 22 related datasets from the 1000-Genome Project, the present study simulates the experiment samples, and extracts the characteristics of biological significance by using classification algorithms in machine learning. The experiment adopts three classification algorithms: Adaboost algorithm, inverse Boosting algorithm and the improved inverse Boosting algorithm, to classify the samples. The generalization error of the three algorithms and the ultimate generalization error of Boosting series algorithms are obtained through experiment. The experiment result shows that the accuracy of detection of gene deletion can reach about 80% using Boosting series algorithms. Consequently, in bioinformatics, the Boosting series algorithms of machine learning can be used to detect gene deletion with better detecting performance.

Keywords: detection of gene deletion; adaboost algorithm; the inverse boosting algorithm; the modified inverse boosting algorithm

1 INTRODUCTION

Bioinformatics is the science which uses computer as a tool for biological information storage, retrieval and analysis. With the rapid development of life science and computer technology, bioinformatics has become one of the most advanced subjects and it has been extensively applied to genetic studies. Among the important genome structural variation types, there is one highly related to some complicated diseases and gene deletion. Therefore, the detection of gene deletion has become one important research subject in bioinformatics and medical field. The existing detection methods of gene deletion include read-pair, split-read[1–2], etc.

With the continuous development of machine learning, its application in biology is also expanding. Examples include the classification of gene expression data[3] in the study of the biological microarray, protein structure prediction[4–6] and protein interaction prediction[7] in proteome research, and gene identification[8] as well as recognition of regulatory elements in non-coding region[9] in genome research. Machine learning research has important significance in detection of genome structural variation. Albers et al.[10] proposed a Bayesian method using short reads data from Illumina sequencing platform to detect genome variation, which effectively improves the detection rate of genome variation detection rate; Grimm et al.[11] introduced Support Vector

Machines (SVM) in biology, by extracting relevant characteristics for SVM training in order to determine whether there is gene variation. In addition, Wittler[12] combined cluster method with biology, and ultimately improved the accuracy in the detection of overlapping variation rate. The above literature demonstrates that machine learning has broad application prospects in gene variation detection. In this paper, a new method to detect gene deletion is proposed using Boosting series algorithms.

One of the tasks in detection of gene deletion is to determine whether there is a deletion in gene sequence. Characteristics of biological significance are extracted from gene data. Some of these characteristics are able to classify the relevant gene data into two groups: with deletion and without deletion. This classification problem can be effectively solved by machine learning. Consequently, this paper uses three Boosting series algorithms of machine learning to address the above mentioned problem. These algorithms include Adaboost algorithm, Inverse Boosting (IB) algorithm[13] and improved Inverse Boosting (IB+) algorithm[14].

2 CHARACTERISTIC EXTRACTION AND EXPERIMENTAL SAMPLES

2.1 Gene deletion analysis

In genetics, gene deletion refers to the fact that gene variation results in the deletion of genetic material from DNA sequence or chromosome. Such deletion can occur in any position and the number of deletion is uncertain[15]. The DNA sequence is composed of several bases (A, G, C, and T). The length of gene fragment represents the number of bases, with the unit bp. In this paper we adopted one popular detection method of gene structural variation, called Paired-End Mapping (PEM) method or read-pair method. This method comprises two fragments and the distance between them. The method uses a large number of individual fragments obtained by high-throughput sequencing technology and maps them to the reference genome using mapping tools based on BWA or BWT algorithms. Then it can be determined whether there is a deletion or not. The detection principle of PEM is as follows. Normally, when read-pair is mapped to the reference genome, the observed mapping direction is invariable and the mapping distance (insert size) is in a specified range according to the insert size in the sequenced library. If the mapping distance is beyond the upper bound of the range, there is a potential deletion (Fig. 1) and the mapping distance equals to the upper bound of the insert size sequence database plus the length of deletion fragment.

2.2 *Characteristic extraction*

Under deletion and non-deletion circumstances, the mapping information around the deletion site reveals differences when the gene fragment is mapped to the reference genome. Therefore, the information on the differences can be used as indicators of the existence of deletion.

The next generation sequencing technology uses a large number of reads of the same length from individual gene, and maps them to the reference genome. In the case of no deletion, the probability of each read mapped to the related site of the reference genome keeps unchanged. In other words, the tested gene and the reference genome have equal number of mapping reads and equal mapping depth (Fig. 2). However, a deletion in the tested gene will reduce the number of reads in the

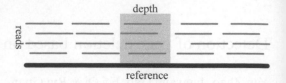

Figure 2. Illustration of the mapping depth in the case of no deletion. The line on the bottom represents the reference genome; the short lines represent the reads, and the number of reads in the shadow area represents the mapping depth in any region of the reference genome.

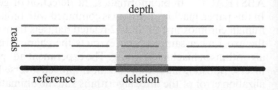

Figure 3. Illustration of the mapping depth in the existence of deletion. The marked area of the reference genome indicates gene deletion, and the number of reads in the shadow area represents the mapping depth at the deletion site.

mapping area and decrease the mapping depth as well (Fig. 3).

By comparison between Figure 1 and Figure 2, it can be noted that the existence of deletion dramatically reduces the number of reads and the mapping depth in the deletion site. Consequently, the number of reads mapped to the deletion site and the mapping depth serve as two important characteristics to determine whether there is a deletion.

The length of read has a certain impact on the mapping result. A smaller length increases the number of mapping sites, and the detection of gene deletion will become more difficult or even lose the mapping significance. However, if the length is too large, the number of mapping sites may be reduced to a certain extent which makes it unable to map. Consequently, the number of meaningful mapping reads can also be used as an important parameter.

The length of the gene deletion fragment and the number of deletion base are various from 1 bp to N bp. The number of reads mapping on the deletion site of longer deletion length is smaller than that of shorter deletion length. As a result, the length of the deletion gene fragment serves as another important parameter.

3 GENERATION OF EXPERIMENTAL SAMPLES

The experimental samples for classification are generated by combining the above characteristic

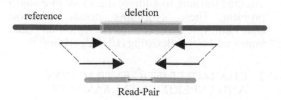

Figure 1. Illustrations of the detection of gene deletion using PEM method. The purple line at the top represents the reference genome; the red shadowed area indicates gene deletion, and the green bars represent read-pair.

Table 1. The detailed information of samples.

Properties	Sample 1	Sample 2	Sample 3	Sample 4	Sample 5
Chrome number	21	22	21	21	22
The number of read-pair	5000000	5000000	5000000	8000000	5000000
Length of read	50	50	50	50	50
Sample size	470	393	470	470	1310
Sequencing depth	11	10	11	20	10
The number of characteristics	13	13	7	13	13
Variation/non-variation	1:1	1:1	1:1	1:1	1:1

parameters with the classes according to the VCF file. The process of generating of the experimental samples consists of data preprocessing and characteristic parameter extraction. Data preprocessing finally generates the standardized SAM format genetic data file. The steps of preprocessing is as follows:

- According to each gene Structural Variation (SV) information in VCF file, the deletion individuals are generated with 50% probability. Thus the ration of variation number to non-variation number is 1:1. The VCF file records the important mapping information, such as the deletion sites, the deletion length and so on.
- Each generated individual corresponds to a Fasta format file, according to which reads file is generated through simulation.
- The reads file is mapped to the reference genome using SAMTOOLS which integrates BWA algorithm. The BAM file recording the mapping information is finally generated. The reference genome used this paper is the 21th and 22th chromosome of the 1000-Genome Project.
- Because the BAM file is binary, the BAM format is translated into the SAM format by SAMTOOLS.

The length of read should be moderate. In the experiment, the length is 50 bp. The sequencing depth is the ratio of the total size of bases to the size of the genome. This depth is one of the indexes to evaluate the sequencing capacity. There is a positive relationship between the sequencing depth and the genome mapping depth. The calculation formula is as follows:

Sequencing Depth

$$= \left(Number_{ReadPair} \times Length_{Read} \times 2 \right) / Length_{Chome}$$

(1)

where $Number_{ReadPair}$ is the number of read-pair; $Length_{Read}$ is the length of read, and $Length_{Chome}$ is the length of chromosome with $chome = 1, 2, 3 ..., 23$.

The experimental samples are generated through the above 4 steps in the format of

$S = \{T_1, T_2, ..., T_n, C\}$, where T_n represents the characteristic parameters with $n = 1, 2, 3, ..., 13$, C indicates the classifications. $C \in \{0,1|1,0\}$, 0,1 and 1,0 represent deletion and non-deletion respectively. According to the different number of read-pair, the size of sample, the sequencing depth and the number of characteristics, the experiment generates 5 samples, whose detailed information is shown in Table 1.

4 THE MACHINE LEARNING MODEL

The Adaboost algorithm, IB algorithm and IB+ algorithm are all subject to the Boosting series algorithms. Using the method of linear integration of several weak classifiers, the Boosting algorithms is able to generate a strong classifier[16]. In this way the weak learning algorithm can be enhanced in order to imporve the learning accuracy.

The Adaboost algorithm is one of the most popular and most widely used Boosting algorithms. It integrates some weak classifiers, whose classification accuracy is merely better than random guess. The core idea of this algorithm is to increase the weights of the correctly classified samples and the weights of those incorrectly classified. The Adaboost algorithm has the over-fitting problem on the erroneous samples or the difficult samples. Considering the problem above, Kuncheva et al. proposed the IB algorithm, the training process of which is contrary to the Adaboost algorithm. The IB algorithm decreases the weights of incorrectly classified samples and increases the weights of the correctly classified ones, so as to relieve the over-fitting problem. Although the IB algorithm partly improves the generalization performance, but the results are still unsatisfactory. Therefore, the IB+ algorithm is proposed.

Based on the inverse sample weight adjustment strategy, the IB+ algorithm integrates a part of subnets to generate the intermediate layer network and uses the intermediate layer network to determine the classification correctness[14]. Suppose that the training sample set is $S = \{(x_1, y_1), ..., (x_m, y_m)\}$,

$y_i \in \{-1, +1\}$ and P represents the weak learner, then the IB+ algorithm can be described as follows:

1. Initialize weight $\omega_i^l = 1/m$, $i = 1, ..., m$; and set the beginning subnet identification of the intermediate layer network $S = l$, where l represents the well trained subnet.
2. For $t = 1, ..., T$:
   (1) Train weak learner using distribution ω_i^l, and get the weak classifier $h_t(x): x \to \{-1, +1\}$.
   (2) Compute the error of $h_t(x)$:

   $$\varepsilon_t = \sum_{i=1}^{m} \omega_i^t [h_t(x_i) \neq y_i].$$

   (3) If $\varepsilon_t > 0.5$ or $\varepsilon_t > p$ and $t < T$, go back to step (2), otherwise compute the subnet weight $\alpha_t = 1/2 \ln 1 - \varepsilon_t / \varepsilon_t$.
   (4) Integrate the subnets from S to t:

   $$h_m(x) = sign\left(\sum_{l=s}^{t} \omega_l h_l(x)\right).$$

   (5) Update the sample's weight:

   $$\omega_i^{t+1} = \frac{\omega_i^t}{Z_t} \times \begin{cases} \dfrac{1}{2(1-\varepsilon_t)} & \text{if } h_t(x_i) = y_i \\ \dfrac{1}{2\varepsilon_t} & \text{if } h_t(x_i) \neq y_i \end{cases},$$

   where Z_t is a normalization factor.

3. Output the final classifier:

$$f(x) = sign\left(\sum_{t=1}^{T} \alpha_t h_t(x)\right).$$

Step (4) represents the important improvement, and step (5) shows the inverse weight strategy.

5 EXPERIMENT AND ANALYSIS

The experiment adopts the three algorithms mentioned above to classify the samples. The number of weak classifiers is 20 and the number of iteration is 500. In the sampling phase, 4/5 samples were randomly selected as training samples and the rest as test samples. Each algorithm was run for 5 times. Then the generalization error of the three algorithms on the 5 experiment samples can be obtained. Further experiment computes the average generalization error of each column using Equation (2). Then the arithmetic mean value of the generalization error of the three algorithms was calculated for every sample and is shown in the last column (AVERAGE) of Table 2. Then we can calculate the Average Generalization Error (AVG) of each algorithm, which is shown in the last row of Table 2. Although the generalization error has fluctuations within a narrow range, it is basically

Table 2. The statistical table of generalization error.

1			2			3			4			5			
ADA	IB	IB+	ADA	IB	IB+	ADA	IB	IB+	ADA	IB	IB+	ADA	IB	IB+	AVERAGE
0.164	0.270	0.245	0.321	0.305	0.359	0.343	0.419	0.381	0.2	0.249	0.274	0.407	0.299	0.413	0.310
0.183	0.232	0.232	0.333	0.297	0.333	0.317	0.370	0.343	0.187	0.223	0.239	0.284	0.309	0.364	0.283
0.177	0.245	0.211	0.313	0.313	0.295	0.281	0.362	0.332	0.151	0.238	0.208	0.250	0.305	0.350	0.269
0.160	0.247	0.221	0.3	0.310	0.295	0.272	0.366	0.326	0.147	0.238	0.218	0.249	0.299	0.347	0.266
0.160	0.251	0.221	0.262	0.313	0.3	0.253	0.383	0.326	0.149	0.253	0.210	0.252	0.313	0.344	0.266
0.158	0.253	0.217	0.256	0.3	0.308	0.260	0.383	0.313	0.143	0.270	0.197	0.241	0.332	0.354	0.266
0.145	0.249	0.215	0.269	0.295	0.308	0.243	0.385	0.302	0.136	0.260	0.197	0.228	0.340	0.364	0.262
0.153	0.266	0.213	0.228	0.295	0.318	0.196	0.385	0.292	0.128	0.260	0.202	0.230	0.339	0.365	0.258
0.143	0.277	0.204	0.223	0.308	0.318	0.202	0.385	0.279	0.145	0.262	0.200	0.227	0.342	0.372	0.259
0.140	0.281	0.211	0.195	0.321	0.333	0.221	0.381	0.277	0.151	0.266	0.205	0.220	0.343	0.365	0.261
0.147	0.289	0.217	0.203	0.331	0.326	0.198	0.383	0.281	0.149	0.270	0.200	0.229	0.341	0.380	0.263
0.143	0.287	0.221	0.210	0.346	0.336	0.215	0.389	0.283	0.140	0.285	0.200	0.218	0.346	0.385	0.267
0.147	0.289	0.223	0.218	0.354	0.341	0.206	0.389	0.285	0.140	0.292	0.205	0.202	0.347	0.398	0.269
0.155	0.287	0.221	0.236	0.356	0.339	0.219	0.389	0.289	0.143	0.294	0.200	0.197	0.346	0.401	0.271
0.147	0.277	0.228	0.210	0.359	0.351	0.238	0.387	0.287	0.151	0.296	0.205	0.197	0.350	0.405	0.273
0.151	0.268	0.226	0.190	0.364	0.356	0.213	0.387	0.292	0.145	0.298	0.205	0.195	0.350	0.402	0.269
0.238	0.251	0.226	0.185	0.364	0.359	0.187	0.392	0.294	0.147	0.298	0.202	0.192	0.350	0.402	0.272
0.217	0.247	0.226	0.182	0.367	0.354	0.199	0.389	0.294	0.143	0.302	0.205	0.197	0.356	0.401	0.272
0.192	0.251	0.223	0.182	0.359	0.356	0.217	0.389	0.302	0.136	0.311	0.210	0.207	0.356	0.405	0.273
0.185	0.247	0.226	0.182	0.367	0.364	0.215	0.387	0.302	0.136	0.315	0.208	0.202	0.352	0.407	0.273
AVG															
0.165	0.263	0.221	0.235	0.331	0.332	0.235	0.385	0.304	0.148	0.274	0.210	0.231	0.336	0.381	

134

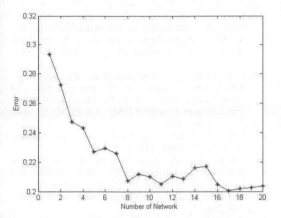

Figure 4. Illustration of the generalization error in the detection of gene deletion based on the Boosting series algorithms.

in the range of 0.2–0.3. The detailed statistics are shown in Table 2.

$$
\begin{cases}
AGE_j = \dfrac{\displaystyle\sum_{i=1}^{K} E_{i_j}}{K}; \\
AVG_{ERROR} = \left(AGE_1, AGE_2, ..., AGE_p, ..., AGE_n\right);
\end{cases}
$$

(2)

In Equation (2) K is the number of operations; E is the generalization error of each operation; AGE represents the arithmetic mean value of the same algoritghm; AVG_{ERROR} expresses the average generalization error, and n is the number of networks.

The data in the last column of Table 2 are used to plot Figure 4 (Fig. 4).

As shown in Figure 4, the curve represents the detection effect of gene deletion based on the Boosting series algorithms. It is evident that when the number of network is 1, the generalization error reaches the greatest value. With the increase in the number of network, the generalization error shows a decline tendency till the number of network grows to 8. Although the generalization error has fluctuations within a narrow range, it is basically in range of 0.2–0.22. When the number of network is 17, the generalization reaches the smallest value of 0.2. Based on the above analysis, we can draw a conclusion that the classification accuracy of the gene deletion can reach 78%–80% using the Boosting series algorithms.

6 CONCLUSIONS

In this paper, the Boosting series algorithms are applied to the detection of gene deletion.

By simulating 5 experimental samples based on the human chromosome 21 and 22 related datasets from the 1000-Genome Project, Adaboost algorithm, IB algorithm and IB+ algorithm are adopted to classify the experimental samples and to obtain the generalization error of the three algorithms, the average generalization error and the final generalization error of the Boosting series algorithms. According to the final generalization error of the Boosting series algorithms, we can draw a conclusion that the classification accuracy in the detection of gene deletion can reach about 80% using the Boosting series algorithms. This conclusion shows that in bioinformatics, the Boosting series algorithms can be used to detect gene deletion and can achieve better detection performance. Because of the large size of gene data and the difficulty in characteristic extraction, the classification time efficiency of the Boosting series algorithms is not satisfactory. Consequently, future research should focus on how to extract more useful characteristics on the premise that the classification accuracy is not affected, so as to reduce the sample data dimension. In addition, we can use some methods like clustering to perform preprocessing on the gene data, and then generate the data sample by choosing the representative gene data from each class. Using the methods mentioned above, the sample size will be reduced.

FUNDING

This research was supported by the National Natural Science Foundation of China (61472026).

REFERENCES

[1] Jin Zhang, Jiayin Wang, Yufeng Wu. An improved approach for accurate and efficient calling of structural variations with low-coverage sequence data[J]. BMC Bioinformatics, 2012, 13(Suppl 6): S6.
[2] Zhang J., Wu Y. SVseq: an approach for detecting exact breakpoints of deletions with low-coverage sequence data[J]. Bioinformatics, 2011, 27(23): 3228–3234.
[3] Golub T.R., Slonim D.K., Tamayo P., et al. Molecular classification of cancer: class discovery and class prediction by gene expression monitoring[J]. Science, 1999, 286(5439): 531–537.
[4] Jones D.T. Protein secondary structure prediction based on position-specific scoring matrices[J]. Journal of molecular biology, 1999, 292(2): 195–202.
[5] Fetrow J.S., Bryant S.H. New programs for protein tertiary structure prediction [J]. Nature Biotechnology, 1993, 11(4): 479–484.
[6] Zhang S.W., Pan Q., Zhang H.C., et al. Classification of protein quaternary structure with support vector machine[J]. Bioinformatics, 2003, 19(18): 2390–2396.

[7] Guo Y., Yu L., Wen Z., et al. Using support vector machine combined with auto covariance to predict protein–protein interactions from protein sequences [J]. Nucleic acids research, 2008, 36(9): 3025–3030.

[8] Korf I. Gene finding in novel genomes [J]. Bmc Bioinformatics, 2004, 5(1): 59.

[9] Bulyk M.L. Computational prediction of transcription-factor binding site locations [J]. Genome biology, 2004, 5(1): 201.

[10] Albers C.A., Lunter G., MacArthur D.G., et al. Dindel: accurate indel calls from short-read data [J]. Genome research, 2011, 21(6): 961–973.

[11] Grimm D., Hagmann J., Koenig D., et al. Accurate indel prediction using paired-end short reads [J]. BMC genomics, 2013, 14(1): 1–10.

[12] Wittler R. Unraveling overlapping deletions by agglomerative clustering [J]. BMC genomics, 2013, 14 (Suppl 1): S12.

[13] Ludmila I. Kuncheva, Christtopher J. Whitaker. Using diversity with three variants of boosting [C]. MCS' 02 Proceedings of the Third International Workshop on Multiple Classifier Systems, 2002: 81–90.

[14] Gao Jingyang, Chenchenglizhao, Zhuqunxiong*. An efficient version of inverse boosting for classification. Transactions of the Institute of Measurement and Control [TIM], Vol.35 No.2, April, 2013, pp 188–199.

[15] Lewis R. Human Genetics: Concepts and Applications [M]. McGraw Hill: New York, 2005.

[16] Tae-Ki An, Moon-Hyun Kim. A New Diverse Adaboost Classifier [C]. The 2010 International Conference Artificial Intelligence and Computer and Computational Intelligence. Sanya China, 2010: 359–364.

Electrical, Control Engineering and Computer Science – Liu (Ed.)
© *2016 Taylor & Francis Group, London, ISBN 978-1-138-02937-8*

Multiple-range query processing in Main Memory Column Store

Hui Liu, Zhijing Liu, Tong Yuan & Jing Wang
School of Computer Science and Technology, Xidian University, Xi'an, China

ABSTRACT: Multiple-range query is one of the most commonly used query types in DBMSs. In this paper, we focus on the processing of such queries in main memory column store, which is a promising physical design for OLAP applications. For this, database cracking is usually adopted in the existing methods. However, under the case of more than one predicates in the query, there comes the challenging problem that the overhead for construction and maintenance of database cracking overshadows the benefit it brought. To solve this problem, a cost model is proposed in this paper to test whether the adaptive index should be used or not in the query plan. Based on this model, we can obtain the optimal query plan for the query. The experimental results show the efficient processing of multiple-range query based on generated query plan in main memory column store.

Keywords: multiple-range query; adaptive index; cost model; main memory column store

1 INTRODUCTION

Multiple-range query processing is one important issue in DBMSs, which applies predicates on more than one attributes. In row store, multi-dimensional structures (e.g. R Tree) are adopted as an index to process such query (Chovanec et al. 2011). With the physical design changes to the Main Memory Column Store (MMCS) (Plattner et al. 2009, Chasseur et al. 2013), in which each column is separately stored and correlated data resided in memory during processing, the traditional structures become inefficient due to multiple access to separately stored columns.

Database cracking (Idreos et al. 2007, 2011, 2012) is usually adopted to realize the multiple-range query processing in MMCS. It is constructed and maintained as a side product of query processing. The core idea of this structure is that data and index structures are reorganized continuously, adaptively, partially, incrementally, and on demand. Thus, the amount of data that will be touched can be significantly decreased, which has a beneficial effect on query processing. Figure 1 shows one example for database cracking. However, in some case, the disadvantage of the adoption of database cracking will outweigh its advantages. For example, when the query has more range predicates, the cost for construction and maintenance of the adaptive index will become unaffordable. So, database cracking is not suitable for all cases. In this paper, we first present several alternative query plans, with adaptive index or not, for the processing of multiple range query in MMCS. Then, a cost model is introduced to

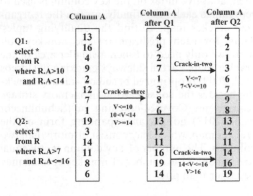

Figure 1. Main principles of database cracking.

measure the execution time of each query plan for the same query. Finally, based on the result of the cost model, we use a heuristic method to choose the most suitable query plan for the current query.

The remainder of this paper is organized as follows. Section 2 introduces related work on this issue. Section 3 describes the cost model and generation of optimal query plan in detail. Section 4 uses a set of experiments to prove the accuracy of the cost model and the efficiency of the generated plan, followed by the conclusion and future work in section 5.

2 RELATED WORK

In this section, we give related work referred to the issue. In MMCS, selection vector (Boncz et al. 2005) is usually adopted to process queries, which realize bulk processing in an efficient way. Vectors, defined as cache-resident and vertical data fragments, are passed through pipelined operators. For multiple range queries, query correlated columns are processed vector-at-a-time and derive intermediate bit vectors. Bitwise anding operator is then applied to intermediate bit vectors of all columns to obtain the final bit vector that indicates which tuples are qualifying all the query predicates. Based on the final bit vector, tuples of user interest can be aggregated or retrieved.

Besides selection vector, processing of such query based on database cracking is also proposed and three approaches have been developed: sideways cracking in (Idreos et al. 2009), partial sideways cracking in (Idreos et al. 2009), and covered-cracking in (Schuhknecht et al. 2013). Basic principle of sideways cracking (Idreos et al. 2009) is given in Figure 1, where M_{PQ} represents cracker map consisting of columns P and Q in relation R. It can be described as follows: first, for n predicates in the query, $n-1$ cracker maps will be constructed with the same key column; then, adaptive index on the key column is used to reorganize each map; finally, we use the reorganized cracker maps to find the qualifying tuples. Thus, for each predicate on the non-key column, we only need to touch a smaller area of the cracker map. Partial sideways cracking in (Idreos et al. 2009) is optimized based on sideways cracking besides the consideration of memory storage restrictions. Covered-cracking in (Schuhknecht et al. 2013) adopts an expanding form of the cracker map, which may consist of more than two columns. The number of key column is still equal to 1, whereas the number of non-key column may be larger than one.

3 QUERY PLAN FOR MULTIPLE-RANGE QUERY

In this section, we present the cost model and generation of optimal query plan. We first describe the space of candidate query plans for multiple-range query processing in MMCS. Then, a cost model is proposed to measure the cost for these plans. Finally, the generation of optimal query plan based on this cost model is presented.

Considering the memory storage restriction, each column in the relation is stored with fixed length values and implicit tuple-IDs. We can calculate the tuple position by the offset to the beginning of the column.

In the cracker map, the key column value and the non-key column value that have the same offset to the beginning of the map belong to the same tuple in the original relation.

3.1 *Candidate query plans*

Based on the usage of database cracking (i.e. use or not, partially use or completely use) in query plans, we can classify them into three types: selection vector based query plan, crack-based query plan, and hybrid query plan. Combined with TPC-H Q6, as Figure 2 shows, we will present a concrete description of the query plans.

Selection vector-based query plan is executed by the adoption of selection vector. First, we set the selection vector size according to the cache size; then, the columns referred to the current query are processed by the granule of the vector size. In each vector, the corresponding query predicate is applied to filter the query tuples, and the intermediate results is represented as a bit vector (1 is defined as qualifying tuples and 0 is defined as opposite). When each current vector of all the query-referred columns is processed, we use bitwise Anding to obtain the final results, namely the tuples qualifying all predicates in the current vector, and is also represented as a bit vector with the same definition. The final bit vector is then applied to retrieve the data that of interest. Figure 3 shows the selection vector-based query plan for TPC-H Q6.

Crack-based query plan performs database cracking on cracker maps, methods described in

select	sum(l_extendedprice*l_discount) as revenue
from	lineitem
where	l_shipdate >= data '[DATE]'
	and l_shipdate< date '[DATE]' + interval '1' year
	and l_discount >=[DISCOUNT] – 0.01
	and l_discount<[DISCOUNT]+0.01
	and l_quality < [QUANTITY]

Figure 2. Definition of TPC-H Q6.

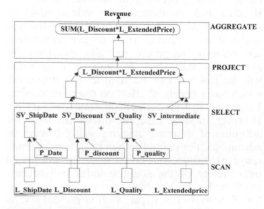

Figure 3. The execution schema of selection vector-based query plan.

138

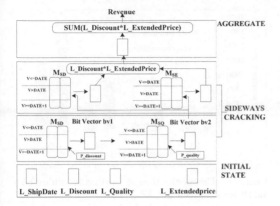

Figure 4. The execution schema of crack-based query plan.

related work belong to this type. Figure 4 shows the crack-based query plan for TPC-H Q6.

As the name shows, hybrid query plan is the combination of the above two methods. In this query plan, some attributes are testified by database cracking and the other by scanning; finally, the result of scanning and the key column in cracker map will be combined as a cracker map. Through the cracker map, we can acquire the tuples that satisfy all predicates in the query.

3.2 Cost model

The above three plans are suitable for different scenarios. So, we need a cost model to estimate the execution time of each query plan for the same query, and then we can generate an optimal query plan with a minimizing cost for the multiple-range query.

During the generation of optimal query plan, it is necessary to determine the access method for predicates on each column in the query, scanning or cracking with the key column. Equation 1 gives the cost for query Q, where m indicates the number of range attributes that the current query referred to and $C_k C_i$ indicates that the cracker map consists of key column C_k and non-key column C_i.

$$Cost(Q) = \sum_{i=1}^{m} \min(\cos t_{scan}(C_i), \cos t_{crack}(C_k C_i)) \quad (1)$$

In the following, we give the cost for scanning and cracking separately. For scanning, its efficiency is mainly determined by the size of the column. Equation 2 gives the cost for scanning on the column, where n indicates the size of column and C_{read} indicates the execution time for reading and testing one tuple.

$$Cost_{scan}(T) = n * C_{read} \quad (2)$$

For cracking, we assume that the key column has been determined. Equation 3 gives the cost for cracking on the cracker map which consists of key column M and non-key column N, where n indicates the size of the data that will be cracked in the cracker map. $C_{exchange}$ indicates the execution time for exchanging two values and is machine dependent. The cost for cracking is mainly determined by the number of exchanging operations. In the best-case situation, namely the data are already in order, the cost will be zero. However, in the worst-case situation, namely the data are in reverse order, all data will be exchanged. In Equation 3, we consider the average case for cracking.

$$Cost_{crack}(MN) = (n/2) * C_{exchange} \quad (3)$$

3.3 Generation of the optimal query plan

We illustrate the generation of optimal query plan based on the above cost model. During the generation, we always choose the execution way that has the minimal cost in each step. We divide the referred attributes in the query into two kinds: attributes with cracker map and attributes without cracker map. For the first kind, it provides the key column that will be used to construct the cracker map for the other attributes. For the second kind, we use the cost model to determine the accessing way: construct cracker map or directly scanning. The final query plan is represent by the tree structure, which indicates the execution sequence of the query plan.

The main steps of this generation are as follows:

Step 1: check the existing cracker maps for the referred attributes in the query. If all exist, select the cracked-based query plan for the query, go to step 6; if not, go to step 2;

Step 2: if a large part of the cracker maps exist, we compare the cost for scanning to the cost for construction of cracker maps. If the scanning cost is larger than construction cost, go to step 3, else, go to step 4; if none of the cracker map exist, go to step 5;

Step 3: construct cracker maps for each attribute without cracker map based on the same key column, then the crack-based query plan is selected, go to step 6;

Step 4: perform scan-based query plan on the attributes without cracker map, and then combine the scanning result with the key column as a new cracker map, then, the hybrid query plan will be chosen, go to step 6;

Step 5: choose the minimal selectivity in all predicates, the corresponding column will be selected as the key column, go to step 3;

Step 6: output the query plan generated for the current query.

The new cracker map, mentioned in step 4, consists of the key column and one bit-vector. Values 1 in the bit-vector represent the tuples that satisfy the predicates on the scanning attributes. The map will be reorganized according to the predicate on the key column. Thus, the result of cracking and scanning can be combined together in the hybrid query plan.

4 EXPERIMENTAL RESULTS

In this section, we give a set of experiments to testify the efficient processing of multiple-range queries. The experiments are implemented on machines running Ubuntu 12 with 16 GB of main memory. These experiments are based on TPC-H benchmark. The variable SF indicates the size of the relation lineitem, which equals to SF*6000000. Because we focus on processing of multiple range queries on one relation, we choose Q6 in TPC-H as the query to be executed, which referred 4 columns in the relation lineitem. The definition of Q6 is given in Figure 2.

We run the workload that consists of 1000 queries in the experiments. All queries have the form of TPC-H Q6, with different random values for the three parameters: DATE, DISCOUNT, and QUALITY. We analyze the experimental result from two aspects: the execution time of each query and the cumulative time of the whole workload.

Figure 5 shows the execution time of each query in the workload with SF = 100. We can see that the execution time of the query sequence based on Selection Vector (SF) is around 1500 ns and with a slight change (150 ns) for each query. For crack-based plan and optimal plan, the average cost of the query sequence is lower but with a higher cost for the first several queries. This is due to that crack-based plan and optimal plan adopt database

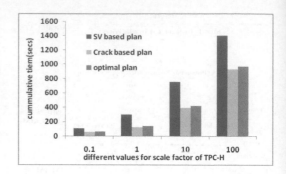

Figure 6. Cumulative time of workloads with different values for scale factor of TPC-H.

cracking totally or partially, which provides a beneficial effect for the subsequent queries.

Meanwhile, compared with crack-based plan, the optimal plan avoids the execution time peak in the query sequence and especially the first several queries, which is obtained by a selection of optimal query plan based on the cost model.

Besides the improved performance for each query in the workload, cumulative time for the whole workload is also a significant indicator. Figure 6 shows cumulative time of the workloads with different values for scale factor of TPC-H. We set SF with four values: 0.1, 1, 10, and 100. We can see that for various sizes of the data, the cumulative time based on optimal plan keeps the same with crack-based plan and lower than the Selection Vector (SV)-based plan.

Consequently, the optimal plan generated based on the cost model can obtain a better performance for both individual query and the whole workload.

5 CONCLUSION AND FUTURE WORK

The existing methods for processing multiple-range queries in MMCS are inefficiency in some cases. To address this problem, a generation of optimal query plan is proposed in this paper, which is based on the estimation of one cost model. The experimental results show that the query plan generated based on the cost model can obtain a good performance.

In the future work, we will focus on the parallelism of the on modern cpu architectures, such as SIMD and mutli-core, to further improve the performance of multiple-range query performance in MMCS.

ACKNOWLEDGEMENT

This work is partly supported by the Technology and Co-ordinate Innovation Project of Shaanxi

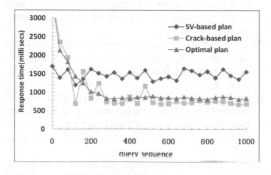

Figure 5. Execution time of each query in the workload.

Province (No.2012 KTZD-02-05-2), the National Natural Science Foundation of China (No. 61202177), and the National Key Technology R&D Project of China (No. 2012BAH01F05).

REFERENCES

Boncz P.A., Zukowski M., Nes N. MonetDB/X100: Hyper-Pipelining Query Execution[C], CIDR. 2005, 5: 225–237.

Chasseur Craig, Patel Jignesh M. Design and evaluation of storage organizations for read optimized main memory databases. Proceedings of the VLDB Endowment, 6 (2013) 1474–1485.

Chovanec P., Krátký M. Processing of Multidimensional Range Query Using SIMD Instructions. Informatics Engineering and Information Science. Springer Berlin Heidelberg, (2011) 223–237.

Idreos S., Stratos, Martin L. Kersten, and Stefan Manegold. Database Cracking. CIDR. 7. (2007) 7–10.

Idreos, Stratos, Martin L. Kersten, and Stefan Manegold. Self-organizing tuple reconstruction in column-stores. Proceedings of the 2009 ACM SIGMOD International Conference on Management of data. ACM, (2009) 297–308.

Idreos S., Stefan Manegold, Kuno Harumi, etc. Merging what's cracked, cracking what's merged: Adaptive indexing in main-memory column-stores. Proceedings of the VLDB Endowment, 4 (2011) 585–597.

Idreos S., Manegold S., etc. Adaptive indexing in modern database kernels. Proceedings of the 15th International Conference on Extending Database Technology. ACM, (2012) 566–569.

Plattner H. A common database approach for OLTP and OLAP using an in-memory column database. Proceedings of the 2009 ACM SIGMOD International Conference on Management of data. ACM, (2009) 1–2.

Schuhknecht F.M., Jindal A., Dittrich J. The Uncracked Pieces in Database Cracking. Proceedings of the VLDB Endowment, 7 (2013) 97–108.

Electrical, Control Engineering and Computer Science – Liu (Ed.)
© 2016 Taylor & Francis Group, London, ISBN 978-1-138-02937-8

A high-efficiency sorting algorithm on multi-core

Xin Huang, Zhijing Liu & Tong Yuan
School of Computer Science and Technology, Xidian University, China

ABSTRACT: Sorting is the most fundamental operation in database system. There are many classical sorting algorithms in the past, but with the development of multi-core SIMD processors, different architectures call for more flexible algorithms to achieve high-efficiency sorting. Therefore, in this paper, we present a sorting algorithm. It cannot only take advantages of SIMD instructions, but makes full use of parallel merge sort as well. Our implementation consists of two phases: an in-core sorting phase and an out-of-core merging phase. The in-core sorting phase implements with 128-bit SSE, and the out-of-core merging phase implements with both odd–even merge and bitonic merge.

Keywords: sorting; mergesort; SIMD; database

1 INTRODUCTION

Sorting is used for numerous computer applications [1]. It is not only a fundamental operation in DBMS, but also a core of other operations, such as index operation or search operation. Thus, many operations in database can take advantages of an efficient sorting algorithm [2].

With the development of modern processors, multi-core SIMD processors provide more and more cores, and each core consists of more and more hardware threads. The Single-Instruction Multiple-Data (SIMD) instructions provided by the processor are not suitable for many classical sorting algorithms, such as quicksort. For example, an SSE instruction can load or store 128 bits of data in a vector register, and the instruction is effective only when the data are aligned on a 128-bit boundary. Hence, we need to find a high-efficiency sorting algorithm which could take advantages of SIMD instructions.

In this paper, we present a sorting algorithm, which can take advantage of the SIMD instructions and run in parallel with multiple threads. The contribution of this paper is as follows:

- We present a high-efficiency sorting algorithm for modern multi-core SIMD processors.
- We make full use of SIMD instructions' benefits in sorting algorithm.
- Our algorithm provides a parallel model of merge sort that performs efficient.

The rest of the paper is organized as follows. Section 2 presents the background and related work. Section 3 details our sorting algorithm on multi-core, including in-core phase and out-of-core phase. Section 4 presents the experimental results. Section 5 contains the conclusion and future work.

2 BACKGROUND AND RELATED WORK

Over the past few decades, a lot of classical sorting algorithms have been proposed. Quicksort is the one of the most widely used algorithms among them, and there are many optimized implementations for quicksort. However, none of them was designed for the modern architectures processors, and it is difficult for quicksort to efficiently exploit SIMD instructions. Thus, a high-efficiency sorting algorithm designed for SIMD processors should be proposed as soon as possible. There were sorting algorithms originally used for sorting on Graphics Processing Units (GPUs), which can be referenced to use in modern processors, as GPUs are generally programmable processors with SIMD instruction sets.

Sorting on GPUs usually uses sorting networks. As a basic algorithm of sorting networks, the bitonic merge sort proposed by Batcher [3] had been widely used. It compares values in a predefined order regardless of the input value, of which we can take advantages to implement the sorting algorithm using SIMD instructions. GPUTeraSort [4], based on bitonic sort, represents data as 2-D arrays or textures to use data parallelism, and hides memory latency by overlapping pointer and fragment processor memory accesses. Furthermore, GPUABiSort [5] was proposed, which is based on adaptive bitonic sort [6] and uses bitonic trees to rearrange the data in order to reduce the number of comparisons. Moreover, the development of GPU capabilities such as flexible comparisons and atomic operations makes it possible for modern GPUs to achieve faster merge sort.

It was Furtak et al [7] who first performed an analysis of exploiting SIMD instructions for sorting small arrays. Their implementation only changed the last few steps of quicksort, and improved the performance of the entire sort. Then AA-sort proposed a multi-core SIMD algorithm based on comb sort and merge sort [8], and it used an odd–even merge network during the merging phase. Later, bitonic merge network can also be used in the merging phase by means of the register shuffle instructions. Our implementation introduces both of them during the merging phase.

3 HIGH-EFFICIENCY SORTING ALGORITHM

In this section, we first introduce two merge networks which is related to our algorithm, then we details our algorithm in two phases: In-core sorting phase and out-of-core merging phase.

3.1 *Merge networks*

There are two merge networks that efficiently exploit the current set of SIMD instructions: odd–even merge network and bitonic merge network. Both of them have multiple steps, and each step executes simultaneous comparisons of elements. Thus, they are suitable for a SIMD implementation.

Figure 1 shows the steps of odd–even merge network for merging arrays of length 4. Array A and array B are held in SIMD registers and both of the arrays in odd–even merge network should be sorted in the same (ascending) order. Odd–even merge network requires fewer comparisons than bitonic merge network. Not all elements are compared with each other, so the overhead of data movement is low.

Figure 2 depicts the bitonic merge network for merging arrays of length 4. Array A and array B are sorted in the opposite order. Thus, if array A is sorted in ascending order, array B should be sorted in descending order. For two sorted arrays in the same order, we need to reverse one of them. Bitonic merge network compares all the elements at every step so that it overwrites SIMD lane at each step. The pattern of comparison is much simpler than the odd–even merge network, and we can also easily implement data movement by using the existing register shuffle instructions, so we choose bitonic merge network in some cases.

Our algorithm exploits both of the merge networks. The in-core algorithm implements with the odd–even merge network, and the out-of-core algorithm uses both the odd–even merge network and the bitonic merge network.

3.2 *In-core sorting phase*

In this subsection, we introduce in-core sorting phase in detail and focus on single thread algorithms.

We first introduce the basic instruction operations in two registers. Here, we use SIMD width of 128 bits. We put 4 elements in each register to vectorize data; thus, we can easily exploit SIMD instruction operations in these two registers, such as compare instruction, swap instruction and transpose instruction.

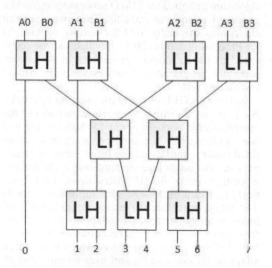

Figure 1. Odd–even merge network.

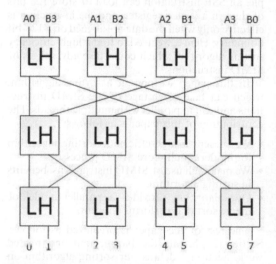

Figure 2. Bitonic merge network.

Figure 3 shows the compare and swap operations in two registers. As shown in the left side, we compare A0 to B0, A1 to B1, A2 to B2, and A3 to B3 without skew, and in the right side, we execute compare operation with skew. Then we swap the smaller one to register A and the larger one to register B.

We usually load k*k numbers into k SIMD registers during the in-core sorting phase and the value k refers to SIMD width. Figure 4 depicts the steps of in-core sorting phase and here we regard each element within a register as a lane, and we first sort the number within each lane by using a series of compare and swap operations. Then, we exploit transpose operation which requires a serious of shuffle operations to let numbers in each register sorted.

The odd–even sorting algorithm used in lane sorting is shown in Figure 5, in which the L stands for low while the H stands for high. Each time we sort a lane with four numbers, and we can obtain a sorted array after five comparisons. For sorting k numbers, it requires $k - 1 + (k(\log k)(\log k - 1))/4$ compare operations.

3.3 Out-of-core merging phase

After in-core sorting phase, we obtain a set of sorted arrays. Then, we load 4 values each from two arrays to merge. When we merge a sorted array length of 4, we choose odd–even merge network. After the first merge, we will have sorted

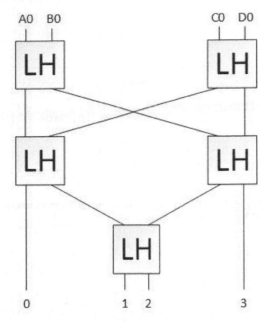

Figure 5. Algorithm used in lane sorting.

arrays length of 8, and we optimize our algorithm using both of the merge networks to merge two sorted array length of 8. As Figure 6 shows, we use bitonic merge network initially to overwrite each SIMD lane. Thus, the input sorted arrays should be in opposite order. As the merging step goes to 4 by 4 merging, we use odd–even algorithm to reduce the data movements, and this algorithm holds for any length of array.

The implementations of mixed merge network requires a serious of shuffle and compare operations. As it shows in Figure 6, the compare operations have an inherent dependency on the shuffle operations that generates the compare latency, and here we use simultaneously merging multiple arrays to overcome it. Since merge operations for different arrays are independent with each other, we can eliminate the latency by interleaving compare operations and shuffle operations for different arrays. It means that we can parallely execute the shuffle operations for one array and the compare operations for another one.

For large data sizes, bandwidth and core numbers may cause poor performance, but with the development of multi-core processors, it is bandwidth not number of cores that becomes the major bottleneck. We introduce multi-way merging to solve the problem. First, we divide the input data into chunks and use our algorithm to sort each chunk. Then, we exploit a binary tree where the leaves point to a sorted chunk, and the root points to the final sorted output to parallel merge the chunks.

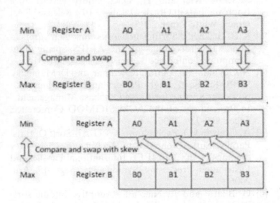

Figure 3. Compare and swap operations between registers.

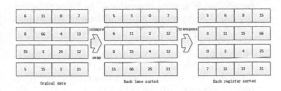

Figure 4. In-core sorting phase (4-wide SIMD).

145

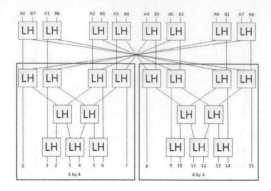

Figure 6. Mixed merge network (merging arrays length of 8).

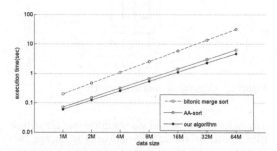

Figure 7. Performance comparisons of bitonic merge sort, AA-sort and our algorithm.

4 EXPERIMENTAL RESULTS

In this section, we show the experimental results of our sorting algorithm on multi-core platform. We run our experiments on Intel multi-core platform with Sandy Bridge architecture. The processor runs at 2.6 GHz and has eight processor cores with two threads per core. Each core has a 32 KB L1 data cache and a 256 KB L2 data cache. The four cores use a shared 20 MB L3 cache with a cache line size of 64 bytes. We compare our algorithm with AA-sort and bitonic merge sort and sort data of different size to evaluate our algorithm. Compared with the traditional sorting algorithm, the novel algorithm not only overwrites each SIMD lane while merging but also controls compare operations and data movements within acceptable limits. The experimental results validate the efficiency of the proposed algorithm in Figure 7.

5 CONCLUSIONS AND FUTURE WORK

In this paper, we proposed a high-efficiency sorting algorithm, which can exploit both the SIMD instructions and thread level parallel merge sort. Our implementation makes full use of modern processor architectures and introduces several new methods to improve the efficiency of sorting. These methods include vectorizing data for SIMD to increase the compute performance, splitting workload to multi-cores, using both of the merge networks when merging and parallel merging for large input data size. Thus, our algorithm makes it possible to sort in a more efficient way in modern multi-core processors. So far, we have focused our algorithm on sorting numbers; however, our algorithm remains to be extended to sorting (key, value) pairs in future work.

ACKNOWLEDGEMENTS

This research is partly supported by the National Key Technology R&D Project of China (No. 2012BAH01F05 and No. 2012BAH01F01-4), the Technology and Co-ordinate Innovation Project of Shaanxi Province (No. 2012KTZD-02-05-2), and the National Natural Science Foundation of China (No. 61202177).

REFERENCES

[1] W.A. Martin. "Sorting," ACM Comp Surv. 3(4): 147–174, 1971.

[2] M.V. de Wiel and H. Daer. "Sort Performance Improvements in Oracle Database 10 g Release2," An Oracle White Paper, 2005.

[3] K.E. Batcher. "Sorting networks and their applications," In Proceedings of the AFIPS Spring Joint Computer Conference 32. AFIPS, 1968; 307–314.

[4] N. Govindaraju, J. Gray, R. Kumar, and D. Manocha. "GPUTeraSort: High Performance Graphics Co-processor Sorting for Large Database Management," In Proceedings of the ACM SIGMOD Conference, pages 325–336, 2006.

[5] A. Greß and G. Zachmann. "GPU-ABiSort: Optimal Parallel Sorting on Stream Architectures," In Proceedings of the 20th IEEE International Parallel and Distributed Processing Symposium, page 45, Apr. 2006.

[6] G. Bilardi and A. Nicolau. "Adaptive bitonic sorting: an optimal parallel algorithm for shared-memory machines," SIAM J. Comput. 18(2):216–228, 1989.

[7] Furtak T, Amaral JN, Niewiadomski R. "Using SIMD registers and instructions to enable instruction-level parallelism in sorting algorithms," In Proceedings of the ACM Symposium on Parallelism in Algorithms and Architectures. ACM Press: New York, 2007; 348–357, DOI: 10.1145/1248377.1248436.

[8] Hiroshi Inoue, Takao Moriyama, Hideaki Komatsu and Toshio Nakatani. "A high-performance sorting algorithm for multicore single-instruction multiple-data processors," 2012;42:753–777.

Electrical, Control Engineering and Computer Science – Liu (Ed.)
© 2016 Taylor & Francis Group, London, ISBN 978-1-138-02937-8

A new scheme of OPGW melting ice and its efficiency

Yuqing Lei, Xi Chen & Yang Wang
China Electric Power Research Institute, Qinghe, Beijing, China

Baosu Hou
Grid Hebi Electric Power Supply Company, Henan Province, China

ABSTRACT: In this paper, a kind of melting ice scheme, achieved through combining new Optical-fiber composite overhead Ground Wire (OPGW) and distributed fiber-optic temperature measurement technology, is researched, to solve the problems of melting ice and monitoring temperature of ground wire. First, the paper introduced the basic system architecture of OPGW melting ice scheme, and then is focused on distribution characteristics of melting ice temperature of the new OPGW ground wire, which are related to the field characters of ice-melting temperature and the time characters of the melting ice. Through a comprehensive analysis of the melting ice scheme of OPGW ground wire, we proposed a well-designed scheme combined with distributed optical-fiber temperature measuring technology and embedded OPGW structure, which is good with effectively monitoring the melting ice process of ground wire and with in real-time adjusting melting ice strategies, thus to shorten time of melting ice and to improve efficiency of melting ice.

Keywords: OPGW melting ICE; distribution optical-fiber temperature measurement; temperature of melting ice; efficiency of melting ice

1 INTRODUCTION

The icing of transmission line is very common in some areas of Guizhou, Hunan, Jiangxi, Sichuan provinces, and so on. Due to the special geographical and climatic conditions, for each February–April, transmission lines in the power system usually go through a period of ice-covering. During this period, the power companies in the region have to organize a lot of manpower and resources to closely monitor the status of transmission line icing. Once found there is icing line to reach a critical value, the de-icing work of ground line is needed to start immediately, to avoid the accidences with breaking line and lifting tower due to ice [1–2].

In this paper, combined with distributed optical fiber temperature measurement technique and a new Optical-fiber composite overhead Ground Wire (OPGW) [3–5], a melting ice scheme of ground wire is presented, which is to realize ground wire melting ice by using heating circuits embedded in optical fiber composite overhead ground wires, and can achieve a real-time temperature monitor of the whole overhead ground wires.

2 OPGW ICE-MELTING SCHEME

Ice-melting system combined with embedded OPGW includes a DC ice-melting device, an OPGW cable for temperature measuring, an icing monitoring system, and a fiber optic measurement temperature system. The connecting relationship of the system is shown in Figure 1. According to the construct requirements of the demonstration line, fiber optic measurement temperature system and icing monitoring system are installed in the substation, when startup, an optical signal for measuring icing situation is sent from substation to each icing monitoring measurement point of an transmission line, and after modulated with icing situation, the signal is turned back to the grating host and through the host's calculation, the ice thickness data are resolved out; and with the same mechanism, the optical signal for measuring ice-melting condition from the substation is sent and then

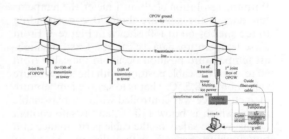

Figure 1. System architecture of OPGW melting ice scheme.

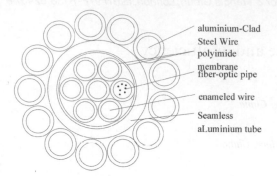

aluminium-Clad
Steel Wire
polyimide
membrane
fiber-optic pipe

enameled wire

Seamless
aLuminium tube

Figure 2.   The cross-section of embedded OPGW.

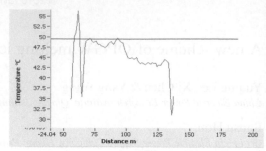

Figure 3.   Distribution temperature curve of OPGW.

carrying the temperature information of the cable returning to the demodulation hosts in substation, to obtain temperature data of the cable. After data analysis, the ice thickness of fiber optic cable, the fiber optic temperature state can be known, and then the melting ice source can be started or stopped, by melting ice control device, and melting ice current can be increased or reduced and so on.

The cross-section of the embedded OPGW used for melting ice is shown in Figure 2. In its design, common ground line is used as reference, while in the middle of the line lies an aluminum tube, in which heat wire and optical unit of stainless steel are embedded. Outside of the tube, many aluminum clad steel wire are roiled in left or right way. The mechanical and electrical properties of this OPGW cable are consistent with the reference traditional ground one. The same as OPGW cable, the embedded OPGW cable, is carrying power system communication functions, as well as fiber optic sensing measurements.

3   TEMPERATURE CHARACTERS OF EMBEDDED OPGW

3.1   *Temperature field characters of embedded OPGW*

Figure 3 is a temperature curve about the fiber optic cable of 70 meters, collected with the distributed optical fiber temperature measurement system. With the resolution of about 1 meter, the temperature data of the cable are about 70, corresponding to the sites of 60 to 130, shown in Figure 3. From these temperature data we can see: First, the highest temperature position occurs in the cable from 60 to 95 of the cable position, with the temperature value about 48.5 °C. That is to say the temperature of melting ice was controlled within a reasonable range, which is below 110 °C, and is safe enough to ensure the safety of the cable performance and functionality. Second, within distance range of the cable noted 60–95, the average temperature is

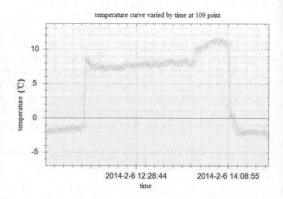

Figure 4.   Temperature curve of OPGW in melting ice.

approximately 48.5 °, while within distance range of the cable from 95 to 130, the average temperature of the cable is about 43.5 degrees.

This phenomenon shows that the transmission line's temperature distribution is very complex, due to the impact of micro-meteorological environment, the impact of wire's fixed fittings, as well as a slight difference between the wire materials. All of them may lead to the local temperature different, and then the temperature of wire is not uniform. And this provides enough instruction that monitoring in real time is necessary for transmission line condition, and distributed temperature measurement technology can truly achieve the actual operating conditions, and it is one of the better means of monitoring safe operation of the lines.

3.2   *Temperature characters with time of embedded OPGW*

Figure 4 data are collected from the ice melting process with the new OPGW cable in Hunan Xuefeng Mountain natural icing test site (station). Fiber optic temperature monitoring data change with time. As can be seen from the temperature curve, the process of melting ice is conditions at an

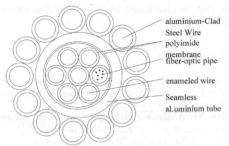

(a) OPGW-24B1-122[117.0; 39.7]

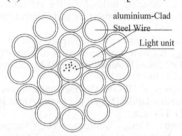

(b)OPGW-12B1+2A1a-68[83.8; 21.7]

Figure 5. Embedded OPGW and traditional OPGW used in trials. (Note: Embedded OPGW cable: OPGW-24B1-122 [117.0; 39.7], wire cross-sectional area: 121.6 mm², DC resistance: 1.708 Ω/km, outer diameter: 14.6 mm, single weight: 837 kg/km. The wire structure is shown in Figure 5(a). Conditional OPGW cable: OPGW-12B1 + 2 A1a-68 [83.8; 21.7], wire cross-sectional area: 68 mm², DC resistance: 1.264 Ω/km, outer diameter: 11.4 mm, single weight: 485 kg/km. The wire structure is shown in Figure 5(b)).

ambient temperature of –5 °C, and the wind speed less than 8 m/s. Temperature of melting ice process of the OPGW ground conductor is at 5 °C ~ 8 °C, at which continuous melting can be realized. The whole melting ice time is about 150 minutes. Significant differences appear in the cross position of ice melting period and off-icing period about the temperatures of the OPGW ground: when melting ice completed, with other states remain unchanged, there is a significant rise in temperature of the OPGW ground, which is completely consistent with the before melting ice control scheme.

4 EFFICIENCY OF THE EMBEDDED OPGW

Icing test chamber and temperature monitoring system used for ice-melting trials are shown in Figure 7. Through melting ice test platform, we can analyze the relationship of ice-melting time and the currents used between the two type wires under typical operating conditions (temperature, humidity, wind speed, and ice thickness).

In test, the selection of environmental conditions is –5 °C, –10 °C (–8 °C), wind speed selects 1 m/s. Since it was hard to control the ice thickness, and was slightly different each time, they are normalized to easily use for comparison. In Table 1 and Table 2, the normalized data of an embedded and an traditional OPGW cable were listed.

Through the test data listed in Table 1/2 we can see: at the same ambient temperature, wind speed,

Table 1. Test data of embedded OPGW.

Serial	Icing thickness (mm)	Environmental temperature (°C)	Environmental humidity (%RH)	Environmental wind speed (m/s)	Melting ice current (A)	Melting ice time (Min)	Joule—heat generation of melting ice (J/km)
1	15	–5	85%	1 m/s	104	58	64 × 106
2	15	–5	85%	1 m/s	121	37	56 × 106
3	15	–10	85%	1 m/s	120	85	125 × 106
4	15	–10	85%	1 m/s	135	50	93.4 × 106

Table 2. Test data of traditional OPGW.

Serial	Icing thickness (mm)	Environmental temperature (°C)	Environmental humidity (%RH)	Environmental wind speed (m/s)	Melting ice current (A)	Melting ice time (Min)	Joule—heat generation of melting ice (J/km)
1	15	–5	85%	1 m/s	200	102.2	310 × 106
2	15	–5	85%	1 m/s	220	83.6	307 × 106
3	15	–5	85%	1 m/s	254	74.3	363 × 106
4	15	–8	85%	1 m/s	232	118.6	484 × 106
5	15	–8	85%	1 m/s	251	113.5	542 × 106

ice thickness and other working conditions, Joule heat of melting ice of Embedded OPGW cable are different from those of a traditional OPGW cable. For example, at ambient temperature −5 °C, with the same of other conditions, the current used by embedded OPGW cable is small (approximately 100 A level), and the Joule heat being needed is about 50 ~ 70 kJ/m values. But for a traditional OPGW cable, the current used is large (approximately 200 A level), the Joule heat required about 300 ~ 370 kJ/m. This shows that an embedded OPGW cable OPGW cable has a better ice-melting efficiency than ordinary.

5 CONCLUSION

This paper describes a new OPGW cable for melting ice and a fiber optic sensing system for icing monitoring, as well as its scheme for melting ice. And through a large number of experiments in a natural icing testing site, the paper studied a control strategy to guide the process of melting ice by using the changing temperature information. In this melting ice scheme, not only the optic-fiber sensor technology is being used, to improve the process of melting ice monitoring capabilities by viewing the cable temperature in real time, but also the temperature monitoring function and wire melting ice function are combined, so melting ice

time can be shortened by adjusting the temperature of melting ice.

REFERENCES

[1] Ji Jin-chuan Gao Yi-bin. Reason analysis of OPGW breakage caused by ice cover [J]. North china Electric Power, 2008(7): 15–17.
[2] Hu Yi. Analysis and counter measures discussion for large area icing accident on power grid [J]. High Voltage Engineering. 2008, 34(2): 215–219.
[3] Zhao Guo-shuai, Li Xing-yuan, Fu Chuang, et al. Overview of de-icing technology for transmission lines [J]. Power System Protection and Control, 2011, 39(14): 148–154.
[4] Liu Wen-Tao, He Shi-zhi, Chen Yi-ping, et al. Defensive strategy for wide area ice disaster of power grid based on DC deice [J]. Automation of Electric Power System, 2012, 36(11): 102–107.
[5] Liming. Effects of OPGW DC De-icing Process on Communication Optical Fiber [j]. Electric Power ICT. 2013, 11(7): 124–128.
[6] Li Chun-hui, Deng Wei-feng, Xu Chang-zhi. Technical Analysis of Temperature Effects on OPGW Optical Unit [j]. Electric Power ICT. 2013, 11(6): 106–111.
[7] Li Ming, Xie Shu-hong. Effects of high temperature on lifetime of optical fi bers used in electric power telecommunication [J]. Wire & Cable, 2012(4): 3–7.

Electrical, Control Engineering and Computer Science – Liu (Ed.)
© 2016 Taylor & Francis Group, London, ISBN 978-1-138-02937-8

Research on the login system of double authentication based on fingerprint and password

Xiuqing Wang
College of Electronic Information and Automation, Tianjin University of Science and Technology, Tianjin, China

Yang Li & Xing Yuan
School of Information and Electronics, Beijing Institute of Technology, Beijing, China

Chunxia Zhang & Suli Wang
College of Electronic Information and Automation, Tianjin University of Science and Technology, Tianjin, China

ABSTRACT: At present, most of the authentication system is a single authentication mode; the traditional password authentication mode has the hidden trouble of information security. In order to improve the security, this paper designed a login system of double authentication mechanisms based on fingerprint and password. This system, which used Visual C++ as development software, successfully constructed a login and register interface with fingerprint and password, and with modifications and delete functions. On the basis of existent username and correct password, this system compared fingerprints in the form of 1:1. If fingerprint was matched successfully, user could enter the system finally. The simulation results verified the feasibility of this login system.

Keywords: system login; fingerprint identification; fingerprint preprocessing

1 INTRODUCTION

The traditional authentication mode mainly uses authentication method of username and password. There are so many weaknesses in this approach, for instance, being forgotten or cracked. Authentication biometric features, such as face, fingerprint, iris, etc., with its unique features, are widely used (Xiangkui Fan 2010). At present, most of the authentication systems are a single authentication method, In order to improve the security, this paper presents a dual feature authentication scheme based on fingerprint and password. Based on the user name, password authentication mode, the fingerprint identification method applied to the authentication process, is used to further protect the security of the login link. Fingerprint identification, as an authentication technology, uses the unique fingerprint feature of each person to verify user's identity and makes up for the deficiencies of the simple, single user account and password. greatly improving the security level of identity authentication.

2 OVERALL SYSTEM STRUCTURE

In general fingerprint systems, such as fingerprint attendance machine, fingerprint access control, etc., fingerprint matching function is usually completed in the fingerprint acquisition module. If the fingerprints are downloaded to module, the security of the system will be reduced, while storage and computing speed of the module also cannot meet the needs of big data of large enterprises.

The system uses the fingerprint acquisition module TFS-M71(Ten fingers Science and technology 2014) to collect the fingerprint image and obtain a characteristic value, upload fingerprints to the PC platform where fingerprints can be processed, and user information is stored. The fingerprint module adopts a split structure, contains fingerprint processing board and fingerprint sensor. The STM32F205RC high-speed digital processor is the core of the handling board. Fingerprint sensor used high precision optical fingerprint acquisition device TFS-D400, it is sensitive for fingerprint and fast speed of recognition, the false rate is less than 0.001%, FRR <0.1%, and fingerprint input time t < 0.1 seconds. The hardware design of double authentication login system is shown in Figure 1.

Fingerprint module is responsible for collecting fingerprint images and uploading them to the PC for storage. Fingerprint processing and comparison modules of the user interface are responsible for fingerprint image preprocessing, feature extraction, and matching. The operation of the

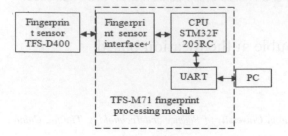

Figure 1. Hardware design diagram of login system.

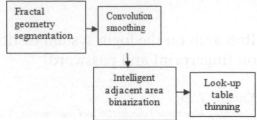

Figure 2. The flow chart of preprocessing.

user interface realizes the storage of user name, password and fingerprint features, as well as, when implementing function of login or deleting, the interface will retrieve information of the database.

3 FINGERPRINT ACQUISITION

This system acquired fingerprint images by module TFS-M71, and transferred them to PC through the USB-TTL interface wire. When the PC has sent capturing command via serial port to the module, TFS-M71 module collects user fingerprint three times, analyzes the common features of three fingerprint images, carries on the logical operation to three fingerprint information, extracts the common features as the characteristics of the final fingerprint information, and then uploads it to pc for storage. Host sends a command of capturing image for the first time to fingerprint module. Complete the second and third collection model after the first time in turn.

4 FINGERPRINT PROCESSING

In order to compare two fingerprints simply and accurately, the processing of images is a very important component. Fingerprint image processing mainly includes three parts, namely preprocessing, feature extraction, and feature comparison.

4.1 Preprocessing

Because of environmental factors having impact on the image acquisition, fingerprint image is usually with much noise. In order to accurately extract fingerprint features, first of all, the collected grayscale fingerprint image should be preprocessed (Hailing Liu 2013, Rajan Amin, Max Bramer & Richard Emslie 2003). The preprocessing flowchart is shown in Figure 2. Fingerprint image segmentation is separation of the background from the fingerprint object, and deletes the background image. In order to remove the image noise, there is a need

Figure 3. Fingerprint image of preprocessing.

to reference surrounding pixels, using the related pixel average template of the object fingerprint to smooth it. And, through binarization, fingerprint images are turned into black-and-white images. In the fingerprint image, the fingerprint characteristic value exists in the form of feature point; therefore, it is necessary to further refinement processing of the fingerprint image.

According to the flowchart, the preprocessing results of segmentation, smoothing, binarization, and thinning are successively shown in Figure 3.

4.2 Feature extraction

Fingerprint includes general and detail features, namely center, triangulation point, endpoint, fork point, etc. The core points are located in the gradual center of fingerprint texture. It is often used for reference when the fingerprint is read and compared to the fingerprint (Amjad Rehman & Tanzila Saba 2014). The core points and triangle points are singular points. Extracting the singular points of the fingerprint according to the principle of the directional field, if a point around the direction field change is fiercer, the greater the possibility of singular point. The grid computing method is used to extract singular points. In detail, one point is the center, along which two closed curves are formed from the counterclockwise direction in 5×5 and 3×3 squares, respectively. If the difference of direction field of two curves is equal, then this point is a singular point. Considering the effect of noise, the pseudo fingerprint feature points are removed. Assemble the endpoint and fork points into the feature point topology, obtaining the fingerprints details.

152

4.3 Feature matching

In fingerprint matching, we first need to carry on the registration of fingerprint image; after registration, fingerprint image matching is completed with the comparison of feature points (Peizhuo Liu 2011). The adopt special point method is the flexible registration method, i.e. to determine the polar axis of the fingerprint image, and the relative rotation and translation by the feature points, to complete the alignment of two fingerprint image.

Fingerprint matching is to compare all feature points of fingerprint template stored in the database with those of the input image. Before the fingerprint feature comparison, there is a need to classify the extracted feature points. Similar feature points constitute the corresponding branch topology. Then, the branch topology is compared. If the angle error and the distance error of template and input feature points are within the upper limit, the two feature points are matched (H. Li & X. Fu 2011). The score of matching feature points is accumulated. If that score is within the acceptable threshold, the fingerprint images match successfully.

5 DATABASE DEVELOPMENT

This system selects ADO technology to realize the interaction with the database ACCESS. ADO is a set of dynamic link libraries that must be imported and initialized before using. By creating connection object instance and calling Open function, the user interface could ultimately realize the connection with database (Jie Tian, Yuliang He & Hong Chen 2005). Through using the Execute function to perform the INSERT and SELETE sentence, the interface completed the operation of storage and query of user data.

The paper designs a type list of users named userstyle and a information sheet of Users named userinfo in the database. The userstyle is used to store the type of users, including administrators and general users. The userinfo is used to store the basic information of users, such as the username, the password, and the fingerprint information.

In the application, by the cooperation of ADO and SQL statements, it is able to complete a series of operations of Access database, such as the realization of database query and data aggregation, and so on.

6 SIMULATION

Double authenticated login interface based on Visual C++ is shown in Figure 4. The operation

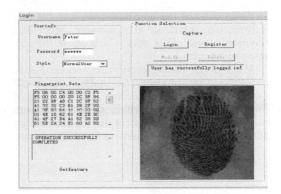

Figure 4. Design of login interface.

interface is with registration, login, modify, and delete function; only the administrator has permission to delete the fingerprint information. The specific process of user login is as follows.

Capture button is responsible for the initialization of the fingerprint module and uploading fingerprint image.

Get feature button is responsible for extracting fingerprint features, and transforming features into a string with size of 196 bytes according to certain rules.

Clicking Login button, we can access user information from the database, which is corresponding with input username in the interface. If the password is matched unsuccessfully, a message box will directly pop-up with words 'wrong username or password, login failed'. On the contrary, if successful, the fingerprints will be compared in the form of 1:1.

Fingerprint matching similarity threshold is to determine the indicator whether the two fingerprints match, with increasing fingerprint threshold, fingerprint recognition error probability becomes large, FRR becomes large, the wrong fingerprint identification and the correct fingerprints probability become smaller, and recognized false rate becomes smaller. Through the threshold adjustment can make the FRR and false accept rate to achieve the best. According to the experiment experience, the similarity threshold of two fingerprints is set at 50. If the sum of matching features' score is less than that value, the two fingerprints are not matched. If fingerprints are matched, the static text box will prompt that user has successfully logged in. At the same time, edit box will show the corresponding user type.

When modifying the user information in the user information interface, the message processing function of the modify button will update the content of the corresponding fields according to the input in the edit box.

When deleting the user information in the user information interface, the message processing function of the delete button will delete the corresponding record.

Multi-group testing experiments showed that: the fingerprint module run stably, and could quickly complete the acquisition and uploading of fingerprint. The login interface interacted well with user database and could accurately realize identity authentication based on password and fingerprint.

7 CONCLUSION

This paper designed a login system of double authentication mechanisms based on fingerprint and password. It is validated by computer simulation that this login system can realize the identity authentication efficiently and accurately. The fingerprint recognition system can complete the user interface and access database by ADO technology, through the operation of user interface to achieve a username, password, fingerprint, and store those data then access them to register, login, modify and delete, query the database information. Based on the user name and password login authentication, combining with the fingerprint identification authentication, the login user identity has secondary certifications. At the same time, with its complex texture, feature values information, it makes up for the deficiencies of the simple, single user account and password authentication, and improves the security of the identity authentication. In addition, through the design and debugging, this system could be used into practical application ultimately.

REFERENCES

Amjad Rehman, Tanzila Saba, "Neural networks for document image preprocessing: state of the ar", Artificial Intelligence Review, Vol. 42, No. 2, 2014, pp.253–273.

Hailing Liu, "Adaptive Gradient-Based and Anisotropic Diffusion Equation Filtering Algorithm for Microscopic Image Preprocessing". Journal of Signal and Information Processing, Vol. 4, No. 1, 2013, pp.82–87.

Jie Tian, Yuliang He, Hong Chen, et al, "A fingerprint identification algorithm by clustering similarity", Science in China Series F: Information Sciences, Vol. 48, No. 4, 2005, pp.437–451.

Li H., X. Fu, "Internet of Things: Algorithm and realization of Fingerprint recognition system: Visual C++", first ed., Posts and Telecom Press, Beijing, 2011.

Peizhuo Liu, "The design and research of the embedded fingerprint recognition system" [D], Beijing University of Chemical Technology, 2011.

Rajan Amin, Max Bramer and Richard Emslie, "Intelligent data analysis for conservation: experiments with rhino horn fingerprint identification" [J]. Knowledge-Based Systems, Vol. 16, No. 5, 2003, pp.329–336.

Ten fingers Science and technology. TFS-M71 Optical fingerprint development module user manual [EB/OL], 2014.

Xiangkui Fan, "Research and application of fingerprint identification in intelligent access control system" [D]. sichuan: Sichuan Normal University, 2010.

Electrical, Control Engineering and Computer Science – Liu (Ed.)
© 2016 Taylor & Francis Group, London, ISBN 978-1-138-02937-8

Efficient adaptive merging indexing for multi-core CPUs

Tong Yuan, Zhijing Liu, Hui Liu & Xin Huang
School of Computer Science and Technology, Xidian University, Xi'an, China

ABSTRACT: Adaptive indexing optimizes the index by reducing the costs of index creation. Unlike traditional full index creation which is performed before any query execution, adaptive index creation is performed as a product of query execution. Recently, adaptive indexing has received much attention, and many variations have been proposed, including database cracking, adaptive merging indexing, and hybrid adaptive indexing. In this paper, we implement adaptive merging indexing for Multi-core CPUs with exploiting thread-level parallelism and data-level parallelism. We first introduce the parallelizing sorting algorithms used in our implementation, including merge sorting and radix sorting. Then, we propose our adaptive merging indexing algorithm. Through experiments on 8-core platform, we finally show that our proposed algorithm has good query performance in memory storage on multi-core CPUs.

Keywords: database index; adaptive merging indexing; multi-core CPUs

1   INTRODUCTION

In a relational database with hundreds of tables and columns, retrieving data can benefit from the use of index when many queries are performed. Thus, indexing is a crucial factor in database system. Too many indexes cause high creation costs and update costs; too few indexes make queries scan more tuples, increasing retrieval costs. The main weakness of traditional indexing is that it covers all rows of a table, even though some rows are rarely or never used. It takes unnecessary time and space to create index on these rows. In order to overcome this weakness, several researchers proposed adaptive indexing techniques (Idreos et al. 2007, Graefe et al. 2010, Idreos et al. 2011), which have received a considerable amount of attention. Adaptive indexing techniques create index dynamically as a product of query execution. The adaptive indexing techniques ensure that index is created and optimized for these tables, columns, and key ranges with actual query execution. The more frequently a key range is queried, the better index is created for it. The key range that is not queried is not indexed.

Meanwhile, since multi-core CPUs have experienced tremendous development, modern CPUs often have four or more cores, and each core has two or more threads. Recently, IBM introduced its new generation processor-power 8, which has 12 cores with 8 hardware threads per core for a total of 96 threads of parallel execution. It has been proved that multi-core CPUs have a bright prospect in the current DBMSs. There are many researches on adaptive indexing, including database cracking (Idreos et al. 2007), adaptive merging indexing (Graefe et al. 2010) as well as hybrid adaptive indexing (Idreos et al. 2011). However, the work (Graefe et al. 2012, Graefe et al. 2014) with respect to database cracking is the only work that is performed on multi-core CPUs.

In this paper, we implement adaptive merging indexing for Multi-core CPUs with exploiting thread-level parallelism and data-level parallelism. We also focus on parallelizing sorting algorithms and avoiding the conflicts with concurrent threads. Experiments on read-only queries prove good performance on our efficient adaptive merging indexing.

2   BACKGROUND AND RELATED WORK

The principle of adaptive indexing is to create and refine index continuously during query execution. Below, we introduce two main adaptive indexing algorithms.

2.1   *Database cracking*

Database cracking pioneered the adaptive indexing, and is the most popular one due to its robustness and simplicity. Database cracking splits the input array into refined partitions as the indexes, and the partition step is similar to the quicksort algorithm where each query brings pivots. Figure 1 shows an example of database cracking. The input data are first copied to index array without sorting.

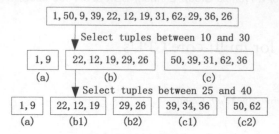

Figure 1. Database cracking.

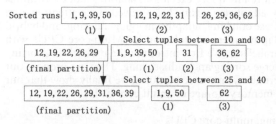

Figure 2. Adaptive merging.

When a query on the key range (10, 30) arrives, the index array is split into three partitions: (1) keys less than 10; (2) keys between 10 and 30; and (3) keys greater than 30. Then a query (25, 40) is performed, thus, more partitions are created. In this example, partitions less than 10, (10, 25), (25, 30), (30, 40) and greater than 40 are created in this step. As the number of queries increases, partitions become smaller and indexes become better optimized.

2.2 Adaptive merging

Adaptive merging also optimizes indexes during query execution. Adaptive merging indexing, which is similar to the merge sorting, first produces equally sized sorted runs. When a query is given, tuples in the key range of query from each run will be merged into a final sorted partition, and other tuples not in that range will be left in their initial runs. Figure 2 shows an example of adaptive merging. When a query on the key range (10, 30) arrives, tuples in that range (10, 30) from initial runs are merged into a sorted partition. Then, a query (25, 40) is performed, we just need to pick out tuples in the range (30, 40) and merge them into the final partition. As with database cracking, the more queries are given, the greater fully indexes are optimized.

3 ADAPTIVE MERGING INDEXING FOR MULTI-CORE CPUS

Adaptive merging indexing proposed by us aims to create and refine indexes efficiently with exploiting

thread-level parallelism and data-level parallelism. Adaptive merging indexing has two main parts: index creation and index optimization. In the index creation, we aim to produce equally sized sorted runs, benefit from multiple cores and SIMD. In the index optimization, we aim to optimize indexes in parallel, including probing tuples in initial runs and integrating tuples into the final partition in parallel.

3.1 Index creation

We first evenly divide the input data into blocks of size M. Then, each thread works on a block to produce a sorted run, which consists of two phases. Here, we adopt the sort strategy similar to the work (Chhugani et al. 2008).

Phase 1: In this phase, we use the sorting network to produce many sorted sequences. The length of the sequence depended on the width of SIMD register. Figure 3 shows the sorting network for four items. We can see that a four-item sorting needs 10 min/max operations and three level comparisons. The beauty of SIMD is that we can handle more sorting operations in parallel, and produce more sorted sequences one time. Figure 4 shows that we exploit data-level parallelism to produce four sorted sequences one time, each of which contains four items. Moreover, transposition operation performed by SIMD shuffle instruction is needed to move items to the correct positions for the next comparison.

Phase 2: After producing many short sorted sequences, we merge them into a sorted run. Merging operation also benefits from SIMD speedup. Figure 5 illustrates the method for merging two sequences of four items. Bitonic merge combines two pre-sorted sequences, A and B, but one is in

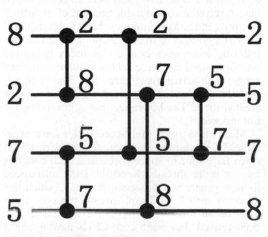

Figure 3. Sorting network.

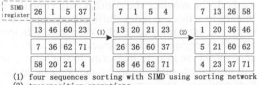

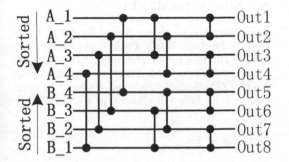

(1) four sequences sorting with SIMD using sorting network
(2) transposition operations

Figure 4. SIMD-register sorting with width 4.

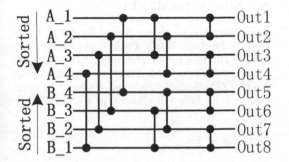

Figure 5. Bitonic merge network.

ascending order, and the other is in descending order. After three level comparisons, a sequence of eight items is created. Through multiple iterations described above, a sorted run is generated. Thus, when threads create a sort run on each block, the index creation is accomplished.

3.2 Index optimization

Let q = (Q$_l$, Q$_h$) be a query with Q$_l$ < Q$_h$, n be the number of sorted runs, and k be the number of threads. The main idea of adaptive merging indexing optimization is very simple. When a query q is given, we will face three situations. (1) If both Q$_l$ and Q$_h$ are in the range of final partition which is created by previous queries, we only need to probe the required tuples in the final partition. This process is similar to the binary search. One thread is performed for the low threshold Q$_l$, and another thread is performed for the high threshold Q$_h$. We do not need to perform insert operation to the final partition in this situation. (2) If neither Q$_l$ nor Q$_h$ locates in the range of final partition, we will perform the probe operation, and then the insert operation. In the probe operation, the thread i works on sorted runs R$_{i+0*k}$, R$_{i+1*k}$, R$_{i+2*k}$, etc., to probe the tuples in the range (Q$_l$, Q$_h$), and to insert them into the final sorted partition. The probe phase can be performed with threads concurrently, because threads work on different sorted runs. In the insert operation, we adopt the Most Significant Digit radix sort. We first create 2^b buckets, where b is the number of bits of the sort digits. Then the thread inserts tuples

probed in its assigned runs into the final partition in parallel, which may incur thread conflicts. To avoid the conflict with concurrent threads, we can adopt several methods such as lock method, independent output method, move-count method, and parallel buffer method, which have been proposed in our previous work (Yuan et al. 2014). Which method to choose mainly depends on the number of buckets, the number of threads, and the input data. Finally, we sort the tuples in each bucket concurrently, as with the sorted run generation. (3) If either Q$_l$ or Q$_h$ is in the range of final partition, we will perform the operation as described in the situation 1 for the required tuples in the final partitions, while we will perform the operation as described in the situation 2 for the required tuples out of the final partitions,

In this way, adaptive merging index is created and optimized on multi-core CPU during the query execution.

4 EVALUATION

In this section, we show the experimental results of our adaptive merging indexing on multi-core platform. We run our experiments on Intel multi-core platform with Sandy Bridge architecture. The processor runs at 2.6 GHz and has eight processor cores that support SMT with two threads per core. Each core has a 32 KB L1 data cache and a 256 KB L2 data cache. The four cores use a shared 20 MB L3 cache with a cache line size of 64 bytes. We use a main-memory column-oriented dataset, represented as an array of fixed size pair <id, value>.

Our dataset has 100 million tuples, and each tuple comprises 4-byte id and 4-byte value. The value is generated at the interval [0,2^{32}) random. We perform 10000 queries on our dataset, and each one follows this type: SELECT SUM (A) from column C WHERE Q$_l$ < A < Q$_h$. Each query will filter the data on the column C, and perform a sum operation over the result. We choose a random value on the low boundary Q$_l$, and the selectivity of query is 1%. We assume that each query is independent from other queries, and each query does not benefit from previous query result. We perform ten times measurements, and show the average result in the figure. Figure 6 shows the accumulated query response time over 10000 random queries, not the time to answer a query.

Moving from left to right of Figure 6, we can see that the query time increases significantly when the query sequence is fewer than 10000, especially when it is fewer than 100. This is due to the reason that the index is being created and refined during the first hundred queries. After 1000 queries, the majority of index has been created and refined, so the time of every query remains stable.

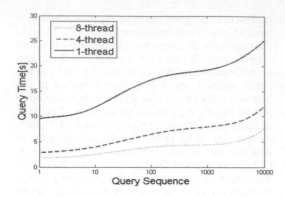

Figure 6. Accumulated query time of our indexing algorithm.

Table 1. Speedup achievement with different numbers of threads.

Threads	Total time	Speedup
1	25.29	1X
4	11.37	2.224X
8	7.43	3.403X

Overall, the experiment results confirm that our efficient adaptive merging indexing has good performance on the multi-core CPUs. From Table 1, we can see the speedup achievement with different numbers of threads. When the number of threads increase to 4, our algorithm achieves a speedup only to 2.224X, while it ideally should achieve 4X. The main reasons are as follows: (1) cache miss on the L3 data cache. If the size of L3 cache is larger, the performance will be better. (2) memory bandwidth. The more threads work, the more contentions for memory bandwidth between threads.

5 SUMMARY

In this paper, we implement adaptive merging indexing for Multi-core CPUs. In the index creation, we use the SIMD and multiples threads to accelerate runs generation. In the index optimization, we utilize parallel radix sort and some methods for avoiding the conflict with concurrent threads to refine the index during the query execution. The experimental results show that efficient adaptive merging indexing achieves good performance on the multi-core platform. In the future, we will implement other adaptive indexing on multi-core platform, and compare them to find out the suitable method for different datasets.

ACKNOWLEDGEMENTS

This research is partly supported by the National Key Technology R&D Project of China (No. 2012BAH01F05,), the Technology and Co-ordinate Innovation Project of Shaanxi Province (No. 2012KTZD-02-05-2), and the National Natural Science Foundation of China (No. 61202177).

REFERENCES

Chhugani, J., A.D. Nguyen, V.W. Lee, W. Macy, M. Hagog, Y-K. Chen, A. Baransi, S. Kumar, and P. Dubey, "Efficient implementation of sorting on multi-core SIMD CPU architecture," Proceedings of Very Large Database, Vol. 1, No. 2, 2008, pp. 1313–1324.

Graefe, G., and H. Kuno, "Self-selecting, self-tuning, incrementally optimized indexes," EDBT International Conference, pages 371–381, 2010.

Graefe, G., F. Halim, S. Idreos, H. Kuno, and S. Manegold. "Concurrency control for adaptive indexing," Proceedings of Very Large Database, Vol. 5, No. 7, 2012, pp. 656–667.

Graefe, G., F. Halim, S. Idreos, H. Kuno, S. Manegold, and B. Seeger. Transactional support for adaptive indexing. The VLDB Journal, pages 1–26, 2014.

Idreos, S., M.L. Kersten, and S. Manegold, "Database cracking," Conference on Innovative Data Systems Research, pages 68–78, 2007.

Idreos, S., S. Manegold, H. Kuno, and G. Graefe, "Merging what's cracked, cracking what's merged: Adaptive indexing in main-memory column-stores," Proceedings of Very Large Database, Vol. 4, No. 9, 2011, pp. 585–597.

Yuan Tong, Liu Zhijing, Liu Hui, and Wang Zi. Hash Partitioning Optimizations Based on MapReduce for Chip Multiprocessors. Journal of Xi'an Jiaotong University, Vol. 48, No. 11, 2014, pp. 97–102.

Electrical, Control Engineering and Computer Science – Liu (Ed.)
© 2016 Taylor & Francis Group, London, ISBN 978-1-138-02937-8

Two-dimensional Compressed Sensing for IR-UWB Wireless Sensor Network data

Yulin Liu
China Electronic System Engineering Corporation, Shenyang, Liaoning, China

Kai Wang, Bo Zhang & Shun Xu
Chongqing Communication College, Chongqing, China

ABSTRACT: To reduce the energy consumption and sampling rate requirement in data transfer between sensor nodes, a two-dimensional measurements based Compressed Sensing (CS) method is proposed for Impulse Radio Ultra-Wideband Wireless Sensor Network (IR-UWB WSN) in this paper. Taking both spatial and temporal correlations into consideration, the CS measurement model is developed based on block quasi-Toeplitz structured matrix and the IR-UWB WSN data is measured in both spatial and time dimensions. Simulation result demonstrates that IR-UWB WSN data recovery by the proposed approach achieves significantly saving of transport cost and sampling rate with small reconstruction error.

Keywords: compressed sensing; wireless sensor networks; IR-UWB; measurement; sparsity

1 INTRODUCTION

Impulse Radio Ultra-Wideband (IR-UWB) has been regarded as an attractive communication solution to provide high bandwidth for Wireless Sensor Networks (WSN) to transfer large amount of data collected by sensor nodes, and this makes IR-UWB based WSN (IR-UWB WSN) a hot topic in WSN research [1]. Nevertheless, not only does the data transfer in IR-UWB WSN suffer the problem of energy limitation like all WSN, but it also faces the new challenge introduced by the short duration pulses in IR-UWB communication: extremely high speed Analog-to-Digital Converter (ADC) components are needed, which is difficult to meet under the current level of the hardware chips.

Compressed Sensing (CS) [2–4] has received a great deal of attention in recent years, because it deals with sparse signal reconstruction problems with far fewer samples than required by the Nyquist rate. Previous research showed that CS provided ideal solutions for the two challenges respectively in different fields of research: considering the spatial correlation of nodes in WSN, CS approach can be used to measure data acquired by individual sensors. Only the measurements are reported to the Sink node and the initial data are reconstructed by CS algorithm, leading a reduction of energy consumed in data transmission [5, 6]; when the temporal correlated (sparsity) in IR-UWB systems is considered, CS approach can also be adopted to

recover the sub-sampled IR-UWB signals, leading a reduction of the ADC resources [7, 8]. However, most of the traditional CS focus on single dimension measurement, while in IR-UWB WSN, efficient data transfer needs to adopt the CS approach in both spatial and time dimensions.

In this paper, we propose a CS approach based on two-dimensional measurements for efficient data transfer in IR-UWB WSN. Considering the spatial and temporal correlations, the IR-UWB WSN data is measured by a new block quasi-Toeplitz matrix in both spatial and time dimensions. Then the measurements, which far less than initial data, are transferred and the initial data of all sensors are reconstructed from the measurement at a distant user by CS reconstruction algorithm. Compared to other CS method for network data, our method not only achieves further savings in transport cost with small reconstruction error, but also reduce the dependence on the high speed ADC components.

2 TWO-DIMENSIONAL MEASUREMENT

We consider a randomly distributed grid of sensors that sensing some physical data (e.g. temperature, pressure) and wirelessly transmit them to a central base station. Sensors with wireless transmitter communicate sensor readings simultaneously to Sink. We assume that the sensor nodes and Sink are both implemented in a UWB channel environment and their position is fixed through the sampling and reconstruction process. The Sink collects network data from sensors and transmits them to user which far away from sensor field via Internet or Satellite.

We denote the data sensed at K-th node by $\mathbf{x_k} \in R^N$ $(k \in \{1, 2, ..., K\})$. $x_k(n)$ is used to denote sample n $(n \in \{1, 2, ..., N\})$ in $\mathbf{x_k}$, and assume that there exists a known sparse basis Ψ for R^N in which the $\mathbf{x_k}$ can be sparsely represented.

Consider the n-th $(\forall n \in \{1, 2, ..., N\})$ snapshot of the network data. Each of the sensors simultaneously records a single reading $x_k(n)$ and modulates the reading as $x_k(n)r_k(j)$, where Gauss pseudorandom sequence $r_k(j)$ is generated by K-th node, $j = 1, 2, ..., J$ $(J \ll K)$. Then the J numbers are transmitted in an analog and synchronized fashion to the collection point, obtaining J measurements

$$y(j) = \sum_{k=1}^{K} x_k(n)r_k(j) \tag{1}$$

The collection point automatically aggregates the measurement vector of n-th snapshot data

$$[y_n(1), y_n(2), ..., y_n(J)]^T = \Phi [x_1(n), x_2(n), ..., x_k(n)]^T \tag{2}$$

After J transmission steps, where the measurement matrix

$$\Phi = \begin{bmatrix} r_1(1) & r_2(1) & \cdots & r_{K-1}(1) & r_K(1) \\ r_1(2) & r_2(2) & \cdots & r_{K-1}(2) & r_K(2) \\ \vdots & \vdots & \cdots & \vdots & \vdots \\ r_1(J) & r_2(J) & \cdots & r_{K-1}(J) & r_K(J) \end{bmatrix}_{J \times K}$$

We will emphasize random independently and identically distributed (i.i.d.) Gaussian matrices Φ in this paper, but other schemes are possible, including random ± 1 Bernoulli/Rademacher matrices, and so on.

We reorder the N measurement vectors for all of the snap shots as $\tilde{\mathbf{y}}_j = [y_1(j), y_2(j), ..., y_N(j)]^T$, which can be regarded as data collected by J nodes, and each of the nodes sense a length N signal vector. Transmitting this length N signal vector in UWB channel environment lead the receive measurement vector

$$\mathbf{z}_j = \mathbf{H} \cdot \tilde{\mathbf{y}}_j. \tag{3}$$

where the transfer matrix is $\mathbf{H}$ a "partial" Toeplitz matrix with elements draw from the UWB channel impulse response $\{h_i\}$. We assume that a low-rate ADC is used at the Sink node to collect small number of elements of $\mathbf{y}$. Note that the low

sampling rate at the receiver leads to the changing of the transfer model: new measurements $\mathbf{y}'$ are formed by extracting parts of the entries from y and the transfer model (3) becomes

$$\mathbf{z}_j = [z_j(1), z_j(2), ..., z_j(M)]^T = D \downarrow (\mathbf{H}\tilde{\mathbf{y}}_j) = \mathbf{H}' \cdot \tilde{\mathbf{y}}_j. \tag{4}$$

where $D \downarrow$ denotes a sub-sampling factor of q. The new transfer matrix $\mathbf{H}'$ of quasi-Toeplitz types formed by extracting one row in every q rows from $\mathbf{H}$.

Note that

$$\mathbf{z_j}(m) = \sum_{i=1}^{N} \mathbf{H}'_{m,i} \cdot \Phi \cdot \mathbf{x}_i(k) \tag{5}$$

where $\mathbf{z_j}(1) = [z_1(1) \quad z_2(1) \quad \cdots \quad z_J(1)]^T, j = 1, 2, ..., J$, $m = 1, 2, ..., M$, then the measurement vector after two-dimensional compressed sensing can be written as

$$\mathbf{z} = \mathbf{Ax} \tag{6}$$

where

$$\mathbf{A} = \begin{bmatrix} \mathbf{H}'_{1,1} \cdot \Phi & \mathbf{H}'_{1,2} \cdot \Phi & \cdots & \mathbf{H}'_{1,N-1} \cdot \Phi & \mathbf{H}'_{1,N} \cdot \Phi \\ \mathbf{H}'_{2,1} \cdot \Phi & \mathbf{H}'_{2,2} \cdot \Phi & \cdots & \mathbf{H}'_{2,N-1} \cdot \Phi & \mathbf{H}'_{2,N} \cdot \Phi \\ \vdots & \vdots & \cdots & \vdots & \vdots \\ \mathbf{H}'_{M,1} \cdot \Phi & \mathbf{H}'_{M,1} \cdot \Phi & \cdots & \mathbf{H}'_{M,N-1} \cdot \Phi & \mathbf{H}'_{M,N} \cdot \Phi \end{bmatrix}_{(J \times M) \times (K \times N)}$$

is the block quasi-Toeplitz measurement matrix, the block

$$\mathbf{A}_{i,j} = \mathbf{H}'_{i,j} \cdot \Phi_{J \times K} \tag{7}$$

where the $\Phi_{J \times K}$ is an i.i.d. Gaussian matrix.

Then the reconstruction of IR-UWB WSN data becomes a standard CS problem in which the vector $\mathbf{x} \in R^N$ should be recovered from its two-dimensional measurements $\mathbf{z}$.

3 RECONSTRUCTION OF NETWORK DATA

Based on the analysis in Section II, the reconstruction of IR-UWB WSN data becomes a standard CS problem in which the vector $\mathbf{x} \in R^N$ should be recovered from its two-dimensional measurements $\mathbf{z}$ of the form in (7). According the conclude in [9], the decoder given by

$$\hat{x} = \arg\min \|x\|_1 \quad \text{subject to} \quad z = Ax \tag{8}$$

ensures exact recovery of $\mathbf{x}$ from $\mathbf{z}$.

Linear programming techniques like Basis Pursuit (BP) [10] or greedy algorithms [11] can be used to solve this problem. According to result in [12], the restrictions for the sparsity for a given number on measurements can be described as

$$m \leq CJM/\log KN \qquad (9)$$

where C is a positive constant. Therefore the minimum necessary number of measurements needed for reconstruction can be limited to

$$JM \geq Cm\log KN \qquad (10)$$

4 SIMULATION RESULTS AND ANALYSIS

In this part, we evaluate the performance of the two-dimensional measurements based CS method for network data in terms reconstruction quality. We apply the proposed method to reconstruct the real-world network data gathered by the WSN located in the Intel Berkeley Research lab [13]. Each sensor detected the environmental humidity, temperature and light every 31 seconds for more than one month. We use first 64 data of randomly selected 16 sensors in this WSN for evaluating the proposed method. We use wavelet transform to sparsify both every individual node's data and every snapshot of all nodes' data. The coefficients are then measured by block quasi-Toeplitz measurement matrix and reconstructed using L1-magic [14]. Finally, the initial data in time domain is obtained by inverse transforming of the reconstructed coefficients. We plot the reconstruction result when experiment is done with the measurement is 10% of the data actual size in Figure 1. The reconstruction of transform coefficients and time domain signal are both match the original data very well.

To provide a clearer view of how the reconstruction errors are characterized by the number of measurements, we let the number of measurements $J \times M$ vary from 0 to 512 and the empirical probability of success for each value of $J \times M$ is determined by repeating this process 100 times and calculating the fraction of successes. The reconstruction is declared a success if the data is recovered with the Mean Square Error (MSE) smaller than 0.0002. We plot the empirical probability of success versus the number of measurement in Figure 2. In order to compare the performance of our method to previous CS method for WSN data, we also plot the reconstruction result based on spatial dimensions measurement and time dimension measurement.

Simulation results show that the IR-UWB WSN data can be recovered with the MSE less than 0.0002 when the measurement is 20% of

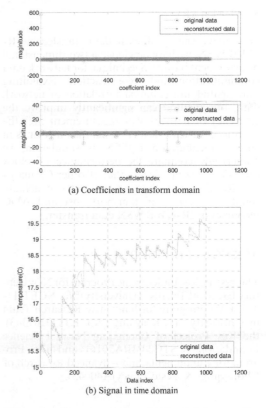

(a) Coefficients in transform domain

(b) Signal in time domain

Figure 1. Reconstructed environmental temperature signal.

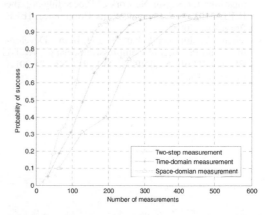

Figure 2. Reconstruction performance for different measurements.

the data actual size. Our method can increase signal reconstruction accuracy with the same number of measurement, or recover the state of the environment with the same quality using less measurement.

5 SUMMARY

To achieve energy efficient data transfer in IR-UWB WSN, we have developed a two-dimensional measurement based CS technique for network data communication and reconstruction. Exploiting both spatial and temporal correlation of network data, our method can significantly improve the efficiency of network data measurement over the traditional scheme, leading a significant savings in energy consumption in WSN. In simulation, we numerically compare the performance of block quasi-Toeplitz matrices to that of other CS matrices. Results show that our method can achieve significant savings in transport cost and ADC recourses in IR-UWB WSN data transfer.

ACKNOWLEDGMENT

This research is funded by the Program for New Century Excellent Talents in University (No. NCET-11-0873), the Program for Innovative Research Team in University of Chongqing (No. KJTD201343), the Key Project of Chongqing Natural Science Foundation (CSTC2011BA2016) and the Program for Fundamental and Advanced Research of Chongqing (No. cstc2013jcyjA40045).

REFERENCES

D. Colling and P. Ciorciari, "Ultra Wideband Communications for Sensor Networks," Proceedings of the IEEE Military Communications Conference, IEEE Press, Oct. 2005.

J. Licheng, Y. Shuyuan, L. Fang and H. Biao, "Development and Prospect of Compressive Sensing", ACTA ELECTRONICA SINICA, Vol. 39, No. 7, 2011, pp. 1651–1662.

D. Donoho, "Compressed sensing," IEEE Transactions on Information Theory, Vol. 52, No. 4, 2006, pp. 1289–1306.

E. Candes and T. Tao, "Near optimal signal recovery from random projections: Universal Encoding strategies", IEEE Transaction on Information Theory, Vol. 52, No. 12, pp. 5406–5425, 2006.

J. Haupt, W.U. Bajwa, M. Rabbat and R. Nowak, "Compressed sensing for networked data," IEEE Signal Processing Magazine, Vol. 25, No. 2, 2008, pp. 92–101.

M. Mahmudimanesh, A. Khelil and N. Suri, "Reordering for Better Compressibility: Efficient Spatial Sampling in Wireless Sensor Networks," Proceedings of IEEE International Conference on Sensor Networks, Ubiquitous, and Trustworthy Computing, IEEE Press, 2010.

L. Yulin, W. Kai and H. Jiwei, "Signal Recovery by Compressed Sensing in IR-UWB Systems," Chinese Journal of Electronics, Vol. 21, pp. 339–344, 2012.

J. Paredes, G. Arce and W. Zhongmin, "Ultra-Wideband Compressed Sensing: Channel Estimation," IEEE Journal of Selected Topics in Signal Processing, Vol. 1, No. 3, 2007, pp. 383–395.

E. Candes, "The restricted isometry property and its implications for compressed sensing," Comptes Rendus Mathematique, Vol. 346, No. 9, 2008, pp. 589–592.

C. Shaobing, D. Donohod and M. Saunders, "Atomic decomposition by basis pursuit," Technical Report 479, Department of Statistics, Stanford University, 2001.

J. Tropp and A. Gubert, "Signal recovery from random measurements via orthogonal matching pursuit," IEEE Transactions on Information Theory, Vol. 53, No. 12, 2007, pp. 4655–4666.

E. Candes, "Compressive sampling," Proceedings of the International Congress of Mathematicians, 2006.

Intel berkeley lab wsn, http://db.csail.mit.edu/labdata.html.

E. Candes, http://www.acm.caltech.edu/.

Electrical, Control Engineering and Computer Science – Liu (Ed.)
© 2016 Taylor & Francis Group, London, ISBN 978-1-138-02937-8

The intermediate classification of hyperplane detection method for clustering SVM

Shangfu Gong, Chang Liang & Xiaoru Bi
School of Computer Science and Technology, Xi'an University of Science and Technology, Xi'an, China

ABSTRACT: Network intrusion detection is an important technology to ensure the security of network. The effective method is utilizing Support Vector Machine (SVM) to realize processing and discrimination of detection data. However, a large amount of data could lead to some defects of SVM method, such as long training time and slow detection speed. This paper proposed a kind of SVM intrusion detection method that is based on intermediate classification hyperplane. First, define boundary face proximity factor for cluster center by the analysis of normal and intrusion detection sample cluster. Then, improve the standard SVM quadratic expression and use cluster center to train for obtaining a closed optimal intermediate classification hyperplane. Lastly, determine the distance threshold for selecting potential support vectors, in order to greatly reduce training samples. The proposed method is experimented on the KDDCUP1999 data-set, compared with the clustering support vector machine method, the experimental results proved that this method can reduce the number of training samples, and effectively improve training and detection speed of SVM.

Keywords: intrusion detection; intermediate classification hyperplane; sample reduction; potential support vectors

1 INTRODUCTION

According to the problem that SVM [1] shows low training and detecting speed and poor real-time in large network data intrusion detection, Zeng [2] adopted the unsupervised clustering algorithm to simplify training samples of SVM intrusion detection. Zhou H applied the k-nearest neighbor method and k-means algorithm to reduce training samples. The above cluster SVM (CLU-SVM) intrusion detection methods complete SVM training with cluster centers rather than the whole training samples. Although improving the training and detecting speed of SVM, due to abandon a considerable number of effective support vectors, it obviously reduces the SVM classification accuracy.

A method based on middle classification hyperplane for SVM intrusion detection is proposed in the paper. Based on clustering normal and attack samples, the method constructs a Middle Classified Hyperplane to reduce the training samples. Experimental results show that the method is effective to reduce the training samples and improve the performance of SVM intrusion detection.

2 SVM INTRUSION DETECTION ALGORITHM BASED ON MIDDLE CLASSIFIED HYPERPLANE

2.1 Defining approaching degree of boundary surface of every clustering centers

Definition 1 If CN_+ and CN_- represent, respectively, the number of cluster centers of normal and attack samples. $D = \{d_{ij}\}_{CN_+ \times CN_-} (1 \le i \le CN_+, 1 \le j \le CN_-)$ is defined as distance matrix of cluster centers. d_{ij} serves as Euclidean distance between No i normal and No j attack cluster centers.

Definition 2 If result of ordering distance matrix D by column is stored in matrix $Q = \{q_j\}(1 \le j \le CN_-)$, $G^+ = \{g_j^+\}(1 \le j \le CN_-)$ is defined as label matrix of normal class cluster centers. No j component of g_j^+ column vector is the line number in D of No i component of q_j column vector. Similarly, if result of ordering distance matrix D by row is stored in matrix $P = \{p_i\}(1 \le i \le CN_+)$, $G^- = \{g_i^-\}(1 \le i \le CN_+)$ is defined as label matrix of cluster centers of attack class. No j component of g_j^- row vector is the column number in D of No i component of p_i row vector.

Definition 3 After clustering center subscript matrix obtained, degree of approaching boundary surface of No i normal cluster center is defined as

$$B_i^+ = (CN_+ - LineNo)/CN_+ \quad i = 1, 2, ..., CN_+ \quad (1)$$

In formula (1), *LineNo* is row number of No i normal cluster center when G^+ is scanned by row; similarly, degree of approaching boundary surface of No j attack cluster center is defined as

$$B_j^- = (CN_- - ColumnNo)/CN_-, \quad j = 1, 2, ..., CN_- \quad (2)$$

In formula (2), *ColumnNo* is column number of No i attack cluster center when G^- is scanned by column.

2.2 *Structure of middle classified hyperplane*

After clustering the normal and attack samples by cluster analysis algorithm [4], the corresponding cluster centers are accessed. And use them to train the standard SVM, then transitional classified hyperplane is got [5,6]. This hyperplane is called the middle classified hyperplane. This construction method equally treats every clustering center and does not consider the importance of each of them. So the similarity degree between middle classified hyperplane and optimal classified hyperplane is reduced. In the paper, approaching degree of boundary surface in every clustering center and its number of samples are put as the cluster center of importance to improve the SVM quadratic expression and generate the SVM, which the weight is importance of samples. The improved SVM is trained with clustering centers to obtain a middle classification hyperplane, so that Possible Support Vectors (PSV) can be obtained more effectively.

Definition 4 In order to reflect the different influences of clustering centers for Middle Classified Hyperplane, normal clustering center of importance degree i is defined as

$$\tau_i^+ = B_i^+ * (Sample\_Num_i^+ / Sample\_Num^+) \quad (3)$$

In formula (3), $Sample\_Num_i^+$ is the samples number of normal cluster i, and $Sample\_Num^+$ is the number of normal samples, $i = 1, ..., C_+$; Similarly attack clustering center of importance degree j is defined as

$$\tau_j^- = B_j^- * (Sample\_Num_j^- / Sample\_Num^-) \quad (4)$$

In formula (4), $Sample\_Num_j^-$ is the samples number of attack cluster j, and $Sample\_Num^-$ is the number of attack samples, $j = 1, ..., C_-$.

Based on the definition above, the form of standard SVM optimization problem is

$$\min_{\omega, b, \varepsilon} 1/2 \omega^T \omega + C \sum_{i=1}^{l} \tau_i \varepsilon_i \quad (5)$$

$$s.t. \quad y_i(\omega^T \phi(x_i) + b) + \varepsilon_i \geq 1, \ \varepsilon_i \geq 0 \ (i = 1, ..., l) \quad (6)$$

The dual problem is

$$\min_{\alpha} Q(\alpha) = \frac{1}{2} \sum_{i=1}^{l} \sum_{j=1}^{l} y_i y_j \alpha_i \alpha_j k(x_i, x_j) - \sum_{i=1}^{l} \alpha_i \quad (7)$$

$$s.t. \sum_{i=1}^{l} y_i \alpha_i = 0 \quad (8)$$

$$0 \leq \alpha_i \leq \tau_i C \ (i = 1, ..., l) \quad (9)$$

Through solving the dual problems, middle classified hyperplane, which is more approximate to optimal hyperplane, can be obtained.

2.3 *Definition of boundary clusters*

Definition 5 If *TrainData* is the training samples set, S represents classified distance, L_i^+ is the distance between any normal cluster center o_i and middle classified hyperplane $g = 0$. Cluster V_i^+ is defined as normal boundary cluster, if L_i^+ meets the condition:

$$L_i^+ - S \leq \lambda^+, \ i = 1, ..., C^+ \quad (10)$$

In formula (10), $\lambda^+ = (|TrainData^+|/|TrainData|) * \eta, 0 < \eta \leq 1, \eta$ is distance threshold; attack boundary clusters can be defined correspondingly.

The Figure 1 shows that most training samples of boundary clusters (marked by black) approaching

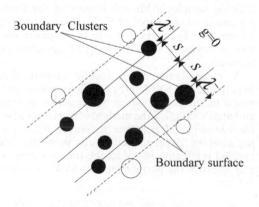

Figure 1. Definition of boundary clusters.

to middle classified hyperplane $g = 0$ are likely to become optimal classified hyperplane $g^* = 0$ possible support vectors. $g = 0$ is largely approximate to classified hyperplane $g^* = 0$, so the samples approaching to $g = 0$ are approaching to $g^* = 0$. Because the samples include by other clusters are far away from $g = 0$ with nothing to $g^* = 0$, they can be deleted from training set, retaining only the one of boundary clusters.

2.4 Selection of possible support vectors

As Figure 1 shows, although boundary clusters are overall approaching to middle classified hyperplane $g = 0$, not all samples included by boundary clusters are approaching to $g = 0$. So it is necessary to reduce the samples.

Definition 6 If the average distance of all normal boundary cluster samples to middle classified hyperplane $g = 0$ is dm^+ and distance of samples $\{x_i, y_i\}$ to dm^+ is v_i^+, close degree of normal boundary cluster samples is defined as

$$CLose\_Degree^+ = 1\Big/\sqrt{\sum_{i=1}^{|N^+|} (v_i^+)^2} \qquad (11)$$

$|N^+|$ represents the number of normal boundary cluster samples. Close degree of attack boundary cluster samples is defined correspondingly.

Definition 7 If S represents classification interval and distance $y_i g(x_i)$ of sample $\{x_i, y_i\}$ of normal boundary clusters to middle classified hyperplane meets the condition:

$$S - \sigma^+ \le y_i g(x_i) \le S + \sigma^+, \ i = 1, \ldots, C^+ \qquad (12)$$

The sample $\{x_i, y_i\}$ is defined as normal possible support vector. $\sigma^+ = Close\_Degree^+ * \alpha$ $(0 < a < 0.1)$ is the selection factor. Similarly, Attack possible support vector is defined correspondingly.

By calculating the distance of every sample of boundary centers and middle classified hyperplane, the samples whose distance is within $(S - \sigma, S + \sigma)$ are retained and others is abandoned. According to formula (12) and Figure 2, this selection strategy deletes not only the samples far away from $g = 0$, but also the overfitting (c,d samples in Figure 2) and noise samples (a,b samples in Figure 2) are removed, which holds over learning of SVM and improves it's generalization ability.

2.5 Description of algorithm

Input: normal training samples TrainData⁺, attack training samples TrainData⁻, clustering number K, selection factor η, selection factor α.

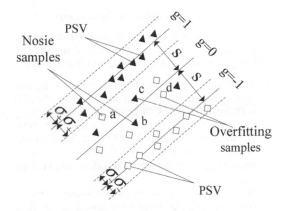

Figure 2. Selection of possible support vectors.

Output: optimal classified hyperplane g^*, and the program code is as follows:

```
CenterSet⁺ = ClusterAnalyze(TrainData⁺,K);
CenterSet⁻ = ClusterAnalyze(TrainData⁻,K);
CenterSet = CenterSet⁺∪CenterSet⁻;
D = DistanceArray(CenterSet);
B = ApproachDegree(D);
g = MiddleClassifyHyperplane(B, CenterSet);
B_Clusters = BoundarySurfaceCluster(g, η);
for i = 1:|B_Clusters|% s(i) is BoundaryCluster
Sample
  If IsPossibleSV(s(1)) = = 1
  ReSample = ReSample∪s(i);
  end
end
g* = SvmTrain(Re_Sample);
```

Main functions in the algorithm are defined as following:

ClusterAnalyzing(TrainData,K): analyzing training samples by clustering algorithm and return cluster centers;

DistanceArray(CenterSet): calculating distance matrix D;

ApproachDegree(D): calculating and return approaching degree of boundary surface in every clustering center;

MiddleClassifyHyperplane(B,C_SampleNum): calculating and return middle classified hyperplane g;

BoundarySurfaceCluster(g, η): obtaining and return boundary cluster set (according to the formula (10));

IsPossibleSV(s(i)): judging whether B_ClusterSample of boundary cluster set is possible SVs according to the formula (12). If so, it returns 1.

SVMTrain(Re_Sample): training SVM with Re_Sample and Return g*.

2.6 Analysis of algorithm time complexity

The time-consuming steps of algorithm are the following several aspects (assume that number of training sample is n, the number of cluster is K):

1. The time complexity of Clustering the normal and attack samples by clustering algorithm [4] is $O(InterTimes * K * n)$;
2. According to reference [7], worst-case time complexity of standard SVM (SMO algorithm) is $O(n^{2.2})$. Therefore, the time complexity of obtaining the middle classified hyperplane $O((2*K)^{2.2})$;
3. When judging boundary cluster samples one by one, the judging times t are based on η, time complexity is $O(f(t))$; and
4. The time complexity of training standard SVM by PSVs (the number is n_{PSV}), is $O((n_{PSV})^{2.2})$. Hence, time complexity of the whole algorithm is $\Theta = O(InterTimes * K * n) + O(f(t)) + O((2*K)^{2.2}) + O((n_{PSV})^{2.2}) \approx O(n)$.

3 EXPERIMENT AND ANALYSIS OF INTRUSION DETECTION

3.1 Experimental dataset and setting of parameters

Two of 10% independent subsets in KDDCUP1999 [8] serves as resource of training set and testing set. 5 experimental datasets are formed after sampling randomly. The RBF function acts as kernel function of SVM and kernel parameter g and C can be obtained by cross-validation optimizing method. The cluster number is K = 10.

3.2 Comparison and analysis of experimental result

Comparing Table 1 with Table 2 under different experimental datasets, the number of reduced training samples (PSV) with MCH algorithm is about 2% of original training samples; therefore, it obviously reduces the training time of SVM. Especially, in the first dataset, training time decreases by 2.057 s. Moreover, this method also effectively reduces of support vectors by 20% on average; hence, detecting time of SVM decreases significantly. Take the fifth dataset for example, detecting time reduces from 4.125 s to 1.563 s. Meanwhile, detecting rate of SVM rises by 0.83% on average and error rate reduces by 0.05%. As a result, performance of SVM is greatly improved.

As illustrated in Figures 3 and 4, compared with CLU-SVM, MCH-SVM shows higher detecting rate and lower error rate in every dataset, especially, in the fourth dataset, detecting rate of MCH-SVM is 18.5% higher than that of CLU-SVM and error rate of MCH-SVM is 0.13% lower than that of CLU-SVM.

In an experiment, we compared CLU-SVM algorithm in training and detecting time of CLU-SVM are greatly reduced, especially, in the third dataset, training and detecting time of MCH-SVM algorithm respectively decreases by 0.01 s and by 1.31 s.

Table 1. SVM intrusion detection results without MCH.

No	Original samples	SV (s)	Training time (s)	Detecting time (s)	Detecting rate (%)	Error rate (%)
1	13740	224	2.094	4.141	93.24	0.25
2	11651	221	2.203	4.047	97.12	0.22
3	14695	220	2.094	3.735	93.47	0.22
4	10980	204	1.765	3.266	91.15	0.2
5	12763	226	2.187	4.125	97.88	0.26

Table 2. SVM intrusion detection results with MCH.

No	Original samples	SV (s)	Training time (s)	Detecting time (s)	Detecting rate (%)	Error rate (%)
1	192	187	0.037	1.370	93.68	0.18
2	185	179	0.087	1.869	98.26	0.20
3	183	180	0.053	1.531	94.22	0.19
4	161	157	0.011	1.100	91.99	0.15
5	191	190	0.056	1.563	98.86	0.20

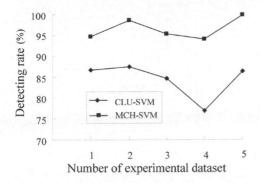

Figure 3. Detecting rate of two algorithms under different datasets.

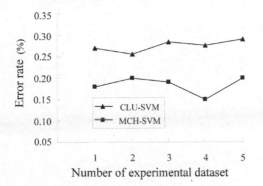

Figure 4. Error rate of two algorithms under different datasets.

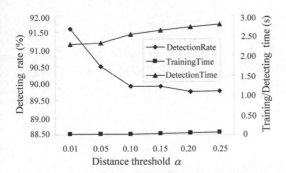

Figure 5. Influence of α on detection performance of MCH-SVM.

As shown in Figure 5, with distance threshold α becoming gradually smaller, training and detecting ime of MCH-SVM not only reduce, but it's detecting rate also rises markablely. This shows that selecting proper distance threshold α not only shrinks the number of training samples but also removes effectively overfitting and noise samples, which can improve raise detection performance of MCH-SVM.

4 CONCLUSION

For the problem of higher dimension and larger network data result in long training time and low detecting speed of SVM, a method for SVM intrusion detection based on middle classification hyperplane is proposed in this paper. Experiments show that the method is effective in decreasing training samples and improving training and detecting speed of SVM. But how to determine the optimal value range of selection factor η boundary clusters will be the next studying content.

REFERENCES

[1] Vapnik V. Statistical Learning Theory [M]. New York, USA: Springer, 1998.
[2] Zeng Z-q, Gao J, Zhu S-z. Network Intrusion Detection Model Based on Simplified SVM [J]. Computer Engineering, 2009, 35(17):132–134.
[3] Qian Q, Wang T-H, Zhan R. Relative Network Entropy based Clustering Algorithm for Intrusion Detection [J]. International Journal of Network Security, 2013, 1:16–22.
[4] Zhou X, Wang F-h, Fu J, Lin Jiayang, JIN Z-j. Mechanical Condition Monitoring of On-load Tap Changers Based on Chaos Theory and K-means Clustering Method [J]. Proceedings of the CSEE, 2015, 6.
[5] Chen N, Xu Z-s, Xia M-m. Hierarchical hesitant fuzzy K-means clustering algorithm [J]. Applied Mathematics A Journal of Chinese Universities (B), 2014, 1.
[6] Liang X-m, Zhou C-y. Speaker Identification Based on Sub-clusters Reduced SVM [J], Computer Engineering and Application, 2010, 46(6):157–159.
[7] Wang L, Sun S-x. Block-based Incremental Training Algorithm of SVM for Very Large Dataset [J]. Application Research of Computers, 2008, 25(1): 98–100+113.
[8] KDD99Cupdataset[DB/OL].http://kdd.ics.uci.edu/databases/kddcup99/kddcup99.Html.

Electrical, Control Engineering and Computer Science – Liu (Ed.)
© 2016 Taylor & Francis Group, London, ISBN 978-1-138-02937-8

Computer simulation on aerodynamic design of waverider vehicle

Lei Luo & Shuang-lin Gao
Xi'an Research Institute of Hi-Tech, China

ABSTRACT: The configuration of a small waverider vehicle is designed using the wedge angle method in this paper. The lift body created has been optimized by the simplex method with a penalty function. The aerodynamic characteristics of the waverider optimized are investigated by the numerical method. The research results show that the wedge angle method is a highly efficient way to generate waverider vehicle; on the different mach number, there is pressure leaking between the upper and lower surface of lift body, which leads to lateral flow in the spanwise on the precompression plane, and which creates the inhomogeneity of inlet flow field; Adding side skirts on the both sides, which can reduce the lateral flow on the waverider 's precompression plane, it can raise the vehicle lift.

Keywords: waverider; wedge angle method; aerodynamic characteristic; computer simulation

1 INTRODUCTION

In recent years, more and more attention has been paid to hypersonic technology around the world, and the United States, Russia, France, Germany, and other countries have gradually made significant progress in hypersonic flight technology, which has moved from stage to explore the concepts and principles to phase of advanced technology development. Waverider vehicles lift upon the forebody and afterbody, expanding nozzle, and the forebody plays a crucial role as the main lift component (Hampton 2002, A. Clark 2006, M.D. Salas 2007). This article describes a quick method of generating waverider lift body, and a forebody has been designed to meet the actual requirements combined with sophisticated optimization technology.

2 METHOD AND MODEL

With the wedge angle method, a waverider shape needed can be generated neatly based on the Osculating Cone theory (R.P. Starkey & M.J. Lewis 1998, R.P. Starkey & M.J. Lewis 1999). Design process is as follows: First, select the Mach number, and choose the shock number and shock angle for the forebody according to the inlet requirements. In this paper, the Mach number is preliminary chosen as: Ma = 6.0, and the first shock angle of the forebody is $\beta_1 = 13^0$.

Based on the waverear relationship:

$$Ma_1^2 = \frac{Ma^2 + \dfrac{2}{\gamma-1}}{\dfrac{2\gamma}{\gamma-1}Ma^2 \cdot \sin^2\beta - 1} + \frac{\dfrac{2}{\gamma-1}Ma^2\cos^2\beta}{Ma^2 \cdot \sin^2\beta + \dfrac{2}{\gamma-1}}$$

(1)

Calculate the first waverear Mach number Ma_1, and then figure out the second waverear Mach number Ma_2 according to the first waverear Mach number Ma_1 and the second shock angle β_2.

The shock angles can be designed by the uniform shock strength, namely:

$$Ma\sin\beta_1 = Ma_1\sin\beta_2 = Ma_2\sin\beta_3$$

(2)

And, it can also be designed by the uniform shock angle, namely: $\beta_1 = \beta_2 = \beta_3$. The initial of the research uses the latter.

Based on the wavefront Mach number and shock angle, the relationship for the shock angle and the flow turning angle is

$$\tan\delta = \frac{Ma^2\sin^2\beta - 1}{\left[Ma^2\left(\dfrac{\gamma+1}{2} - \sin^2\beta\right) + 1\right]\cdot\tan\beta}$$

(3)

Calculated the flow turning angle of waverear for each shock angle: $\alpha_1, \alpha_2, \alpha_3$.

The selected inlet curve is shown in Figure 1. The inlet form curve can be designed to a rectangular model, and the edges are connected with smooth curves.

The method of the waverider is: In condition of the Mach number and zero angle of attack designed, all the shocks intersect at the lower surface line of the inlet. After the success of the design, the shock wave generated at O_1 point of the forebody intersects to point B, and the second and third shocks generated at O_2 and O_3 points in the same vertical cross-section also intersect at point B. The details are shown in Figure 1. For this purpose, the design is as follows:

1. Arbitrarily select a point O of the upper surface form curve of the inlet, and find the O_3 point along the direction of negative X-axis, to make the angle between OO_3 and X-axis to: $\alpha_1 + \alpha_2 + \alpha_3$, and the angle between O_3B and X-axis to: $\beta_3 + \alpha_1 + \alpha_2$.
2. Set point O_3 as a starting point, and find the point O_2 along the direction of negative X-axis, to make the angle between O_3O_2 and X-axis to: $\alpha_1 + \alpha_2$, and the angle between O_2B and X-axis to: $\beta_2 + \alpha_1$.
3. Set point O_2 as a starting point, and find the point O_1 along the direction of negative X-axis, to make the angle between O_2O_1 and X-axis to: α_1, and the angle between O_1B and X-axis to: β_1.
4. The first leading edge curve of the forebody can be generated by points O_1, which are tracked by the upper surface form curve of the inlet, and the second and the third leading edge curves are generated by point O_2 and point O_3, respectively.
5. The first and the second leading edge curves compose the first surface of the lower forebody surface; the second and the third leading edge curves compose the second surface of the lower forebody surface; the third leading edge curve and the upper surface form curve of the inlet compose the third surface of the lower forebody surface; and all the three surfaces compose the lower surface of waverider forebody.

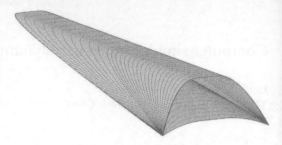

Figure 2. Lift body configuration optimized first.

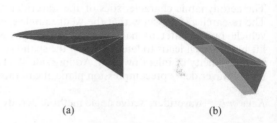

(a)   (b)

Figure 3. Lift body configuration optimized second.

6. The upper surface of waverider forebody is obtained by the translation of the first leading edge curve along the free flow, that is X-axis.

This initial design parameters in the paper are as follows: Mach number is Ma = 6.0, shock angle are $\beta_1 = \beta_2 = \beta_3 = 13°$, the height of the inlet is 30 mm, spanwise length is 100 mm, the lower surface form curve of the inlet is straight line, the middle part of the upper surface form curve is straight line, and both edges are connected with the second exponential curve. Aiming to make the lift-drag ratio to the maximum, volumetric efficiency $\eta_{sp} > 0.1$, and taking the forebody length $L > 10.0$ as constraint conditions, the initial design of the waverider configuration can be optimized preliminary. The results are shown in Figure 2. Reference to a small cruise aircraft shape, design parameters such as volume requirements of the forebody, a secondary shape optimization is done, the design results are shown in Figure 3 (H. Xiao 2003).

3 AERODYNAMIC PERFORMANCE OF FOREBODY

Figure 4 shows the pressure coefficient distribution diagrams for the waverider configuration, which is optimized secondary when the design Mach number Ma = 6.0 and off-design Mach number Ma = 7.0, altitude is 30 km, the flight angle of attack, and side slip are zero.

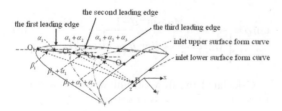

Figure 1. Waverider configuration design principle-based wedge angle method.

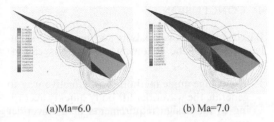

| (a)Ma=6.0 | (b) Ma=7.0 |

Figure 4.   Computational pressure coefficient contours.

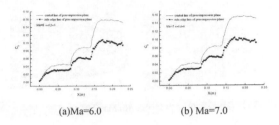

| (a)Ma=6.0 | (b) Ma=7.0 |

Figure 5.   Pressure coefficient variations in constant Ma, α, β.

Comparing the pressure distribution under different Mach numbers shows that the pressure connection exists between the upper and lower surface of the waverider forebody, which will result in the existence of lateral flow in spanwise on the precompression plane. Figure 5 shows the comparison charts of the pressure coefficient of the centerline and edge line on the forebody precompression plane under different Mach numbers. It can be seen from the figure that from the first compression surface to the third compression surface, the pressure leaking is gradually increasing between the upper and lower surfaces, and the pressure coefficient difference of centerline and edge line is gradually increasing too. The existence of these results will cause the non-uniform flow of inlet.

For the shortcomings of the lift body configuration designed, the concept of waverider configuration is used to improve the front design method. By creating successive approximation closed shock, the aim is to close the high-pressure gas in the lower surface, reduce the horizontal flow, and increase the uniformity of inlet flow field.

As the side edge shock can reduce lateral flow, a second side edge and a third side edge in side wall zone are built in this article. The aim is to impede lateral flow by using the shock wave surface, and the diagram is shown in Figure 6.

Figures 7 and 8 show the pressure coefficient distribution under the different Mach numbers, and the comparison chart of the pressure coefficients on the precompression plane central line and edge lines. It can be seen from the figure that the precompression plane centerline pressure coefficient, especially the third precompression plane centerline pressure coefficient, has significantly improved after modified and the range is between 15 and 20%; the range of the edge line pressure is between 20 and 30%, and the gap narrowed significantly. That is, the lower surface flow of forebody modified played a role of reducing lateral overflow, and the pressure of the center line is higher than the pressure of the edge line, still maintaining the effect of boundary layer overflow.

Figure 9(a) shows the static pressure distribution of the lower surface of waverider lift body

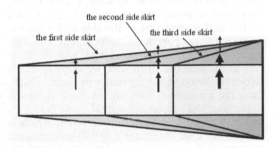

Figure 6.   Lower surface configuration of lift body modified.

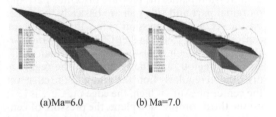

| (a)Ma=6.0 | (b) Ma=7.0 |

Figure 7.   Computational pressure coefficient contours for forebody modified.

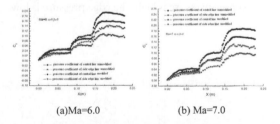

| (a)Ma=6.0 | (b) Ma=7.0 |

Figure 8.   Pressure coefficient variations in constant Ma, α, β for forebody modified.

modified at the design point, while Figure 9(b) shows the static pressure distribution at the off-design point.

As can be seen from Figure 9(a): on the first compression plane, the pressure of center area and

(a)Ma=6.0　　　　　　(b) Ma=7.0

Figure 9.　Static pressure contours for forebody modified.

side wall area are almost the same; on the second compression plane, the pressure of center area is 20% higher than that of the side wall area, the pressure of rear part in the center region is 16.65% higher than that of the front part, and the pressure in the region is distributed regularly from the center to the edge, yet the center pressure is high, the edge pressure is low; on the third compression plane, the pressure of central point is 11.77% higher than that of the marginal point, which is 50% higher than that of the side wall area, and the pressure in the region is also distributed regularly from the center to the edge, yet the center pressure is high, the edge pressure is low.

As can be seen from Figure 9(b): on the first compression plane, the pressure of center part and marginal part and side wall are basically the same; on the second compression plane, the pressure of central point is 13.0% higher than that of the marginal point, which is 30.4% higher than that of the side wall region, and the pressure in the region is distributed regularly from the center to the edge, yet the center pressure is high, the edge pressure is low; on the third compression plane, the pressure of central point is 3.1% higher than that of the marginal point, which is 46.2% higher than that of the side wall region, and the pressure in the region is also distributed regularly from the center to the edge, yet the center pressure is high, the edge pressure is low.

4　CONCLUSION

Based on the above analysis and numerical simulation results, the following conclusions can be obtained:

1. The wedge angle method is an effective way to generate the waverider lift body quickly according to the design requirements of a waverider vehicle.
2. The pressure leaking exists on the upper and lower surface of the lift body optimized, which will result in the existence of lateral flow in spanwise on the precompression plane, causing the non-uniform inlet flow field under the design and off-design state.
3. Adding the side skirts of the lift body at the edge can effectively reduce the lateral air flow on the precompression plane, thus effectively raising the lift of the forebody.

REFERENCES

Clark, A. "Development of an Airframe-Propulsion Integrated Generic Hypersonic Vehicle Model," AIAA 2006–218, 2006.

Hampton, "Research in Hypersonic Airbreathing Propulsion at the NASA Langley Research Center," NASA-TM-2002–0011, 2002.

Salas, M.D. "A Review of Hypersonics Aerodynamics, Aerothermodynamics and Plasmadynamics Activities within NASA's Fundamental Aeronautics Program," AIAA 2007–4264, 2007.

Starkey, R.P. M.J. Lewis, "A Simple Analytical Model for Parameteric Studies of Hypersonic Waveriders," AIAA 1998–1616, 1998.

Starkey, R.P. M.J. Lewis, "Analytical Off-Design L/D Analysis for Hypersonic Waveriders with Planar Shocks," AIAA 1999–3205, 1999.

Xiao, H. "Configuration Design and Optimization of Waveriders," Northwestern Polytechnical University, Xi' an China, 2003.

Communication and computer networks

Electrical, Control Engineering and Computer Science – Liu (Ed.)
© 2016 Taylor & Francis Group, London, ISBN 978-1-138-02937-8

Multiple faults localization theory for transparent all optical networks

Yujie Li
Gansu Electric Power Corporation, Lanzhou, Gansu, China

Xin Li & Yang Wang
China Electric Power Research Institute, Haidian District, Beijing, P.R. China

ABSTRACT: This paper proposes a novel multiple faults localization theory based on the principles of fuzzy clustering theory. The proposed multiple-fault localization theory has high scalability and speed as the vector matching fault-localization.

Keywords: multiple-faults localization; fuzzy clustering theory; all optical networks

1 INTRODUCTION

With the continuous emergence of new network services, such as cloud computing, social media, multimedia streaming, online search, online gaming, the Data Centers (DCs) and Data Center Networks (DNCs) have achieved rapid development. Optical networks have gained tremendous importance due to their ability to support very high data rates using the Dense Wavelength Division Multiplexing (DWDM) technology. At such high rates, a brief service disruption in the operation of the network can result in the loss of a large amount of data. The author implement fuzzy fault set based multi-link faults localization mechanism in multi-domain large capacity optical networks. It has high scalability, speed and success rate compared with extended LVM protocol [1]. The fault may appear at any time, at any place and at any kind with sink node send alarming packet into other sink node along with the nature of fault propagation, the relationship between the fault link and sink node is become uncertain and there is not exist determined value describing the uncertainty (a fault link may belong to one or several sink node at a short period of time temporarily). The fault localization becomes critical, complicated and it is difficult to precisely obtain the number of the fault and identify possible network fault location separately. The LVM protocol is a fault localization protocol for localizing single-link failures in all-optical networks. This protocol assumes no optical power monitoring available at each intermediate node and only an edge node is able to detect the power loss or quality degradation of an optical signal. Therefore, it provides the maximum level of transparency by skipping any power monitoring or spectrum analysis at intermediate nodes on a light path. An advantage of the LVM protocol is that it limits fault localization in a smaller perimeter area and can thus significantly reduce the time and space complexities of fault localization. However, it may take long time to localize a failed link if the affected light paths are long. Therefore, it is necessary to optimize the traffic distribution in a static network in order to minimize the fault localization time at each link. The reader is referred to for more details about the LVM protocol.

2 BASIC PRINCIPLES OF FUZZY THEORY

Fuzzy set theory offers new methods for modeling the inexactness and uncertainty concerning decision making. The fuzzy approach improves the potential for modeling human reasoning and for presenting and utilizing linguistic descriptions (e.g. rather sure) in computerized inference. While we believe that the Fuzzy Clustering Algorithm may be generally applicable in every network scenarios. Thus, we study the performance of Fuzzy Clustering Algorithm in the context of a restricted class of faults for which we can indeed construct practical Fuzzy Clustering Algorithm. In each instance, we collect fault signatures from a fault detection system. The signatures are then input to a localization algorithm that outputs a hypothesis corresponding to a set of likely faults in the network. We use the fuzzy membership function describes the uncertainty between the sink nodes and the fault link based on the network traffic and the network topology. We apply the Fuzzy Clustering Algorithm used in the all-optical transport network, give the steps of Fuzzy Clustering Algorithm for fault localization based on the mathematical model of optical transport network and evaluate the effectiveness of Fuzzy Clustering Algorithm.

A classical (crisp) set includes elements which can either belong to or not belong to the set, whereas elements of a fuzzy set may have various degrees of belonging. A fuzzy set can be defined as follows.

If X is a collection of objects denoted generically by x then a fuzzy set A in X is a set of ordered pairs:

$$A = \{(x, u_A(x)) | x \in X\} \qquad (1)$$

$U_A(x)$ is called the membership function or grade of membership (also degree of compatibility or degree of truth) of x in A. The membership function describes the degree to which the element x belongs to the fuzzy set A [1].

Operations with a fuzzy set are defined via their membership functions. Basic set operations are intersection, union and complement, which are defined as follows:

Intersection: $u_{A \cap B}(x) = \min(u_A(x), u_B(x))$ (2)

Union: $u_{A \cup B}(x) = \max(u_A(x), u_B(x))$ (3)

Complement: $u_{\overline{A}}(x) = 1 - u_A(x)$ (4)

Using these conservative definitions the combined results are determined solely by the maximum in the case of union and the minimum in the case of intersection, independent of the strengths of other inputs [3]. Research has shown that in human reasoning various other operators in the T-norm and Tconorm class or the averaging operators may be better models. In the following only the algebraic sum, which is also a special case of Hamacher-union and Dubois-union, is defined [2].

The a-level set and the cardinality (or "power") of a fuzzy set are also defined in the following due to their use in the proposed method for fault location. The set of elements that belong to the fuzzy set A at least to the degree α is called the α-level set [9]:

$$A_\alpha = \{x \in X \mid u_A(x) \geq \alpha\} \qquad (5)$$

For a finite fuzzy set A the cardinality is defined as

$$|A| = \sum_{x \in X} u_A(x) \qquad (6)$$

The relative cardinality of A is defined as

$$\|A\| = \frac{|A|}{|X|} \qquad (7)$$

In comparing fuzzy sets by their relative cardinality the same universe should be chosen. The relative cardinality can be interpreted as the fraction of elements of X being in A, weighted by their grades of membership in A [2]. In knowledge engineering fuzzy sets support the rule-based reasoning. Expert systems of the first generation utilized rules to model human reasoning. However, the number of rules needed in the knowledge base was huge. In the next generation frames were used to model the knowledge of the application domain and the focus was on model-based reasoning. Fuzzy set theory turns the focus on the experts and their operation.

3 MULTI-FAULTS LOCALIZATION

To help understand the improved LVM protocol, it is necessary to identify different combinations of multiple failures. The FC-LVM protocol is an extension of the LVM protocol. Due to the limitation of space, we will only focus on the difference between LVM and the new requirements for FC-LVM design. For ease of exposition, we will consider the scenarios with two link failures. However, the protocol can also be applied to the scenarios with more than two link failures.

Case 1: De-coupled multi-failure. In this case, two failures occur simultaneously. They are apart from each other and there is no overlapping between the perimeters corresponding to the two failures. Figure 2 gives an example of this case.

Case 2: De-coupled multi-failure shared light path. This case is very similar to Case 1. Two failures occur simultaneously. They are apart from each other and there is no overlapping between the perimeters corresponding to the two failures.

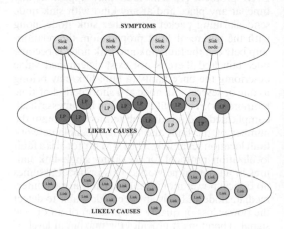

Figure 1. The relationship between the sink nodes, the light path and link.

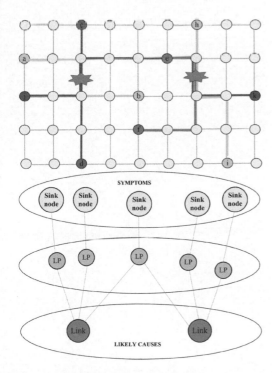

Figure 2. De-couple multi-failure scenario.

However, there is a light path passing through the two failures. Figure 3 gives an example of this case.

Case 3: Special cases. One special scenario is that two failures occur simultaneously and all failed light paths have shared links, as shown in Figure 4. The executive sink node 11 will localize the failed link b and it will assume that no other failed links exist. In this case, one of the failed links will not be localized.

Case 4: Another special scenario is that one failure occurs and all failed light paths have shared links, as shown in Figure 5. In this case, all failed

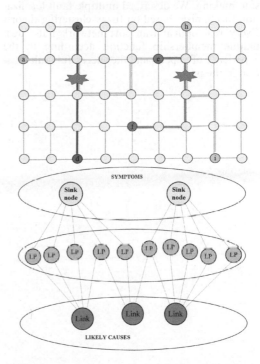

Figure 4. Overlapping perimeter areas with shared light path.

Figure 3. De-coupled multi-failure shared light path.

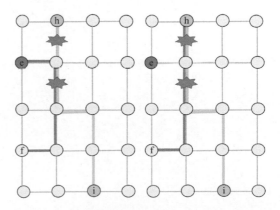

Figure 5. Un-localized failed links.

light paths share two or more links. In our example, all light paths shares link c and b. Therefore, the executive sink node 16 cannot tell exactly which link failed and hence it assumes that both links failed.

4 CONCLUSIONS

Fuzzy set theory offers new methods for modeling the inexactness and uncertainty concerning decision making. We described multiple fault localization mechanism based on fuzzy clustering theory. Every real optical transport network must own unique membership function according to the experience of experts and the successful fault location in the past.

REFERENCES

[1] JingranLuo, Shanguo Huang, Jie Zhang, Xin Li, WanyiGu, A Novel Multi-Fault Localization Mechanism in PCE-Based Multi-domain Large Capacity Optical Transport Networks, OFC, Los Angeles, CA, USA, March 2012.

[2] H. Zeng, A. Vukovic, and C. Huang, "A Novel End-to-End Fault Detection and Localization Protocol for Wavelength-Routed WDM Networks," Proc. SPIE, Photonics North, Toronto, Canada, Sept. 2005.

[3] Limited-perimeter vector matching fault-localization protocol for transparent all-optical communication networks A.V. Sichani and H.T. Mouftah, IET Commun., 2007, 1, (3), pp. 472–478.

[4] Tomsovic K., Ling J.M., A Proposed Fuzzy Information Approach to Power System Security. Third Symposium on Expert Systems Application to Power Systems, Tokyo-Kobe, Japan, April 1991, p. 427–432.

Electrical, Control Engineering and Computer Science – Liu (Ed.)
© 2016 Taylor & Francis Group, London, ISBN 978-1-138-02937-8

Study on characteristics of Channel Plasmon Polariton Waveguides by the improved coupled mode theory

Aning Ma, Yuee Li & Yuzhen Wang
School of Information Science and Engineering, Lanzhou University, Lanzhou, China

Guojian Li
College of Electrical Engineering, Northwest University for Nationalities, Lanzhou, China

ABSTRACT: A classical surface plasmon polariton waveguides is studied by the improved coupled mode theory. By simulating and analyzing the electromagnetic field distribution of Channel Plasmon Polariton (CPP) waveguides at telecom wavelength $\lambda = 1.55 \mu m$, the best structure size of directional couplers (DCs) and communication channel based on CPP waveguides is determined that the channel depth $h = 1.2 \mu m$, angle $\theta = 20°$ and $h = 2.1 \mu m$, $\theta = 20°$. The results obtained reveal that the CPP waveguides is more suitable for used as a communication channel. It is of great value for the practical application of CPP waveguides in integrated optical circuits.

Keywords: Channel Plasmon Polariton Waveguides (CPPW); the improved coupled mode theory for SPPW; communication channel; integrated optical circuits

1 INTRODUCTION

Surface Plasmon Polaritons (SPPs) are prospective methods to accomplish miniaturization and high-density integration of optical circuits [1–3]. A variety of structures have been proposed and experimentally studied to demonstrate sub-wavelength confinement and efficient guiding of plasmon polariton modes including metal stripes [4], V-shaped grooves (CPPs) [5] and triangular wedges (WPPs) [6]. We have proposed an improved coupled mode theory for SPPW in [7], which is different from the coupled mode theory of conventional optical waveguides that satisfies the weakly guiding situations.

In this paper, a comprehensive analysis about CPP waveguides is given. The modal characteristics of CPP waveguides as a function of channel angle and channel depth are given. Two main parameters with different separation distance including coupling lengths and transmission lengths are analyzed. The normalized power in waveguide 1, 2 along CPP waveguide is shown. The results obtained reveal that CPP waveguides is more suitable for serving as a communication channel.

2 PRINCIPLE OF THE IMPROVED CMT

For two symmetrical CPP waveguides with much loss, by simply introducing the imaginary part of propagation constants, the coupled mode equations can be written as

$$\frac{da_1}{dz} = i(\beta + i\alpha)a_1 + i\kappa a_2 \tag{1a}$$

$$\frac{da_2}{dz} = i(\beta + i\alpha)a_2 + i\kappa a_1 \tag{1b}$$

where, $a_j (j = 1, 2)$ is the modal amplitudes of waveguide j; $\beta + i\alpha$ is the complex mode propagation constant of CPP waveguides. β is the phase constant, α is the attenuation constant. In this study, only the metal loss is considered because the dielectric loss is much smaller than the metal loss in CPP waveguides. κ is the mode coupling coefficient which can be calculated as [7]

$$\kappa = \frac{\omega \varepsilon_0}{4} \int\int_{-\infty}^{\infty} \left(n^2 - n_0^2\right) \left[\vec{E}_{1t}^* \cdot \vec{E}_{2t} + \left(\frac{n_0^2}{n^2}\right) \vec{E}_{1z}^* \cdot \vec{E}_{2z}\right] dxdy \tag{2}$$

where, n is the refractive index distribution in the surrounding dielectric media, n_0 is the refractive index distribution of CPP waveguides. E_{jt} and E_{jz} $(j = 1, 2)$ are respectively the transverse electric field distribution and the longitudinal electric field distribution of CPP waveguide j.

Using $a_1 = A_1 e^{-\alpha z} e^{i\beta z}$ and $a_2 = A_2 e^{-\alpha z} e^{i\beta z}$, Eq. (1) is converted to the equations of A_1 and A_2, and supposing optical power P_0 initially launched into waveguide 1, we can obtain the power of waveguide 1, 2

$$P_1 = P_0 \cos^2(\kappa z)e^{-2\alpha z} \quad (3a)$$

$$P_2 = P_0 \sin^2(\kappa z)e^{-2\alpha z} \quad (3b)$$

Then as long as P_0 is given, the power distribution in waveguide 1, 2 along propagation direction z can be obtained by Eq. (3). 100% power coupling occurs at the coupling length of

$$L_c = \frac{\pi}{2\kappa} \quad (4)$$

The transmission length can be calculated by

$$L_t = \frac{1}{2\alpha} \quad (5)$$

Assuming the optical power $P_0 = 1$ initially launched into the first CPP waveguides, we can define the crosstalk as the ratio of the output power in the second CPP waveguides to the input power in the first CPP waveguides. So the crosstalk XT (intrinsic crosstalk) can be calculated as the normalized power in the second waveguide of the distance along the propagation direction

$$XT = P_2(z) = 10\log_{10} P_2(z)/dB \quad (6)$$

3 RESULTS AND INTERPRETATION

The cross section of Au channel plasmon polariton waveguides is shown in Figure 1(a). The radius of curvature of vertex is $r = 10\ nm$. The corners where the flat horizontal surface meets the triangular structure are also rounded with a radius of curvature $R = 100\ nm$. It is worth mentioning that the CPP mode is no longer guided for a height

$h < h_c = 1.0\ \mu m$ [8]. The permittivity of the metal (Au), the cladding (Air) and the substrate (Silica) are $\varepsilon_{Au} = -131.95 + 12.65i$, $\varepsilon_c = 1$ and $\varepsilon_s = 1.47^2$ respectively at telecom wavelength $\lambda = 1.55$. Figure 1(b) presents the distribution of the electric field in the Au wedge plasma waveguides of Figure 1(a).

The mode properties of CPP waveguides as a function of channel angle θ and channel depth h are given in Figure 2(a) and Figure 2(b). In Figure 2(a), the channel angle is varied and the channel depth is constant, $h = 1.2\ \mu m$. It is shown that the best channel angle is $\theta = 20°$, if we consider the influence of normalized effective index and transmission length simultaneously. In Figure 2(b), the channel depth is varied and the channel angle is constant, $\theta = 20°$. It is found that the best channel depth is $h = 2.1\ \mu m$, while the CPP waveguides has

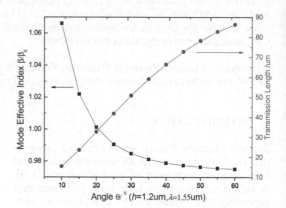

(a)

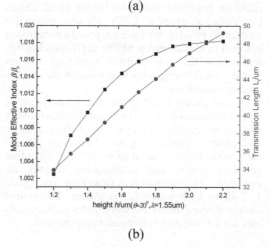

(b)

Figure 2. (a) The modal characteristics of CPP waveguides as a function of channel angle; (b) the modal characteristics of CPP waveguides as a function of channel depth.

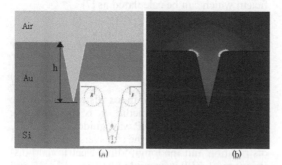

Figure 1. (a) The cross section of Au channel. (b) The distribution of the electric field.

the optimal transmission characteristics, and the best channel depth is $h = 1.2\ \mu m$, while the CPP waveguides has the optimal coupling performance. Therefore, a comprehensive analysis about the transmission and coupling performance of CPP waveguides using the improved coupled mode theory for SPPW is given.

The mode coupling of CPP waveguides is investigated numerically by calculating the mode effective indexes (both real and imaginary parts) of SPP waveguides. The schematics of two adjacent parallel CPP waveguides is depicted in Figure 3. Two CPP waveguides are separated by a distance D. In order to have a meaningful research, we consider exactly two types of geometrical dimensions for CPP waveguides according to the above analysis results. So we can select $h = 2.1\ \mu m$, $\theta = 20°$ and $h = 1.2\ \mu m$, $\theta = 20°$ at $\lambda = 1.55\ \mu m$, with these parameters only the first (fundamental) CPP mode can be guided.

The transmission length $L_t = 1/2\alpha$ and the coupling length $L_c = \pi/2\kappa$ is shown in Table.1. The following conclusions are drawn: (i) The coupling length L_c of adjacent parallel waveguides increases with increasing separation distance D; (ii) The coupling length 166.98 μm is larger than the transmission length 51.096 μm when the channel depth $h = 2.1\ \mu m$ and separation distance is 0 μm; (iii) The coupling length 27.39μm is little smaller than the transmission length 34.54 μm when the channel depth $h = 1.2\ \mu m$ and separation distance is 0 μm; (iv) The channel of depth 2.1 μm is more suitable for serving as a communication channel because of its longer transmission length, and the channel of depth 1.2 μm can be

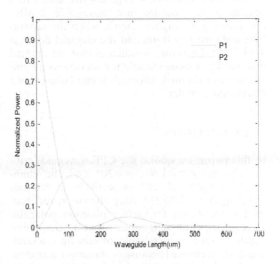

(a) h=2.1μm, $\theta = 20°$

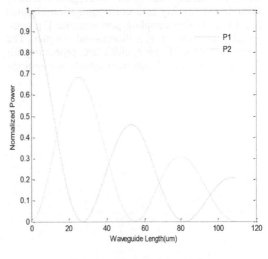

(b) h=1.2μm, $\theta = 20°$

Figure 4. The normalized power in waveguide 1, 2 along CPP waveguide.

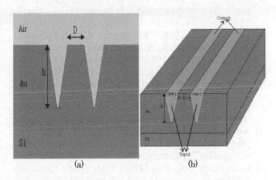

Figure 3. Schematic of two adjacent parallel CPP waveguides. (a) the cross section of two adjacent CPP waveguides; (b) the structure of two adjacent CPP waveguides.

Table 1. The coupling length and transmission length of CPP waveguides.

$h = 2.1\mu m$, $\theta = 20°$	$h = 1.2\mu m$, $\theta = 20°$
$L_t = 51.096\mu m$	$L_t = 34.54\mu m$
$D = 0\mu m$	$D = 0\mu m$
$L_c = 166.98\mu m$	$L_c = 27.39\mu m$
$D = 0.2\mu m$	$D = 0.2\mu m$
$L_c = 274.56\mu m$	$L_c = 80.99\mu m$

constructed a directional couplers because of its smaller coupling length.

More specifically, Figure 4 gives the normalized power exchanges along with the first and second CPP waveguide, when the waveguides separation is $D = 0$ μm. We can see that two CPP waveguides do hardly exchange optical power and the normalized power in the first and second waveguide is decreasing quickly after propagating a certain length because of great attenuation of CPP waveguides, when the separation distance $D = 0$μm and the channel depth is $h = 2.1$ μm in Figure 4 (a). Whereas, in Figure 4 (b), about 70% optical power coupled into waveguide 2 after transmitting a coupling length, when the separation distance $D = 0$ μm and the channel depth is $h = 1.2$ μm. Therefore, it indicates that the channel waveguides is more suitable for serving as a communication channel, although it can be used as a directional coupler.

4 CONCLUSION

In this paper, we studied the CPP waveguide with the channel $h = 2.1$ μm, $\theta = 20°$ and the channel $h = 1.2$ μm, $\theta = 20°$ respectively at telecom wavelength $\lambda = 1.55$ μm using the improved coupled mode theory for surface plasmon polariton waveguides. The results indicate that the channel of depth 2.1μm is more suitable for serving as a communication channel because of its longer transmission length, and the channel of depth 1.2μm can be used for constructing a directional coupler. Therefore, we conclude that the CPP waveguides is more suitable for serving as a communication channel due to its weaker coupling performance [7], compared with using it as a directional coupler. The results about CPP waveguides are promising for the development of sub-wavelength waveguides, high density information storage, highly integrated nano-scale photonic devices and related applications.

ACKNOWLEDGEMENT

This work was financially supported by the Fundamental Research Funds for the Central Universities (lzujbky-2014-48) and National Natural Science Foundation of China (No. 61405083).

REFERENCES

[1] W.L. Bames, A. Dereux, T.W. Ebbesen. "Surface plasmon subwavelength optics." Nature, 424(6950), 2003, pp. 824–830.
[2] S.I. Bozhevolnyi. et al, "Channel plasmon subwavelength waveguide components including interferometers and ring resonators." Nature, 440(7083), 2006, pp. 508–511.
[3] D.K.Gramotnev, S.I. Bozhevolnyi. "Plasmonics beyond the diffraction limit." Nat Photon, 4(2), 2010, pp. 83–91.
[4] Berini, P., "Plasmon polariton modes guided by a metal film of finite width." Optics Letters, 24(15), 1999, pp. 1011–1013.
[5] Pile, D.F.P. and D.K. Gramotnev, "Channel plasmon-polariton in a triangular groove on a metal surface." Optics Letters, 29(10), 2004, pp. 1069–1071.
[6] Moreno, E., S.G. Rodrigo, S.I. Bozhevolnyi, L.M. n-Moreno, and F.J. Garca-Vidal, "Guiding and focusing of electromagnetic fields with wedge plasmon polaritons." Physical Review Letters, 100(2), 2008, pp. 239011–239014.
[7] Aning Ma, Yuee Li and Xiaoping Zhang, "Coupled mode theory for surface plasmon polariton waveguides." Plasmonics, 8(2), 2013, pp. 769–777.
[8] Gramotnev DK, Pile DFP (2004) Single-mode subwavelength waveguide with channel plasmon-polaritons in triangular grooves on a metal surface. Appl Phys Lett 85(26):6323–6325.

Electrical, Control Engineering and Computer Science – Liu (Ed.)
© 2016 Taylor & Francis Group, London, ISBN 978-1-138-02937-8

Interference performance analysis for Device-to-Device system with multi-antenna destination underlying cellular networks

Miaomiao Gao, Shixiang Shao & Jun Sun
Jiangsu Provincial Key Laboratory of Wireless Communication, Nanjing University of Posts and Telecommunications, Nanjing, P.R. China

ABSTRACT: A two-hops wireless communication system with regenerative relays over flat Rayleigh-fading channels is presented in this paper. We consider the system with multiple antennas deployed at the destination node. The cellular user and Device-to-Device (D2D) pair share the same uplink resources. We propose to manage the Cellular Users (CU) interference into D2D system. We derive the outage probability expression and instantaneous Symbol-Error-Rate (SER) upper bound based on the assumption that the perfect CSI is known at all nodes. System simulations show that the proposed algorithms outperform the interfered system with single antenna.

Keywords: Device-to-Device (D2D); decode-and-forward; multiple antennas; interference; outage probability; instantaneous Symbol-Error-Rate (SER)

1 INTRODUCTION

Recently, Device-to-Device communication has been attracting more and more attention by virtue of advantages it offers: increasing the utilization rate, and reducing the load of base station. Introducing relay into D2D communication can provide multi-channel diversity to improve the reliability and enhance the coverage of system. As so far, a lot of works [1] have focused on three-node model. As this is an ever-increasing demand for higher date rate wireless access, so that Multiple-Input Multiple-Output (MIMO) technology is being introduced to the D2D communication to meet this demand. In [2], multiple antennas at the destination node are introduced, which use adaptive Power Allocation (PA) algorithms in order to improve the performance of the system.

In practice part, the interference between CU and D2D has a big effect on the transmission. Allowing D2D transmissions on the same resources as the cellular Downlink (DL) is challenging because of the high mutual interference between the two. One approach is to coordinate the transmissions via interference-aware scheduling [3]. The other is to introduce interference-avoiding preceded Down Link (DL) transmission [4].

In this paper, we reuse the same Uplink (UP) resources at the BS. The interferences cause mostly from CU to D2D receiver. The BS can ignore the interference from the D2D transmitter because of its lower transmit power. We investigate a system, in which the destination node has multiple antennas while source and relay nodes have only one antenna each. We can derive the outage probability expression and instantaneous Symbol-Error-Rate (SER) upper bound with Binary Phase Shift Keying (BPSK) modulation of the system, respectively. Simulation results show that the proposed algorithms have better performance than the interfered system with single antenna. What is more, interference has an effect on the system, so we should not ignore it.

2 SYSTEM MODEL

Considering the cooperative system consists of one source S, one relay R, and one destination D, where multiple antennas can be deployed at the destination node. Here, node S is communication with node D through node R, which acts as relay. We assume that there is no direct link between S and D. We propose a DF-based dual-hop half-duplex relaying system. Let x_s and x_c be scalar symbols transmitted from S and CU, respectively. In the first hop, the received signal from S to R is

$$y_{SR} = \sqrt{p_S}h_{SR}x_S + \sqrt{p}h_{CR}x_C + n_R \qquad (1)$$

where p_S, p is the transmitted power at the source and the CU, h_{SR}, h_{CR} are zero-mean random variables with variance σ_{SR}^2 and σ_{CR}^2, respectively. n_R is the noise at the node R. From eqn. (1), the instantaneous received SNR is

$$r_{SR} = p_S|h_{SR}|^2 / \left(p|h_{CR}|^2 + N_0\right) \quad (2)$$

In the second hop, the received signal from the relay to the destination is given by

$$y_{RD} = \sqrt{p_R}h_{RD1}x_R + \sqrt{p_R}h_{RD2}x_R + \sqrt{ph_{CD}}x_C + n_D \quad (3)$$

x_R is a scalar symbol transmitted from R and p_R is the transmitted power at the relay. h_{RD1} and h_{RD2} denote the instantaneous channel gains of the two paths. n_D denotes the noise received by the destination. Then, the instantaneous received SNR is

$$r_{RD} = \left(p_R|h_{RD1}|^2 + p_R|h_{RD2}|^2\right) / \left(p|h_{CD}|^2 + N_0\right) \quad (4)$$

3 SYSTEM PERFORMANCE

Simply, let $r_s = p_S/N_0$, $r_R = p_R/N_0$, $r = p/N_0$.

3.1 Outage probability

In a DF scheme, the relays successfully decode the message and then forward the detected message to the destination node. If the link quality of any of the two hops falls below a predetermined threshold rth, an outage will occur. With the help of [6], the outage probability of the system is given by

$$\begin{aligned}
P_{out} &= P\{\min(SNR_{SR}, SNR_{RD}) < rth\} \\
&= 1 - P(r_{SR} \geq rth)P(r_{RD} \geq rth) \\
&= 1 - \frac{r_s\delta_{SR}^2 r_R\delta_{RD}^2}{rth^2 r\delta_{CR}^2 + r_s\delta_{SR}^2} \frac{\left(2r\delta_{CR}^2 rth + r_R\delta_{RD}^2\right)}{\left(r\delta_{CR}^2 rth + r_R\delta_{RD}^2\right)^2}
\end{aligned} \quad (5)$$

Proof: See Appendix A.

3.2 Instantaneous SER

Let p_e denote the instantaneous SER for a system transmitting message with SNR r over an Additive White Gaussian Noise (AWGN) channel. Denote $p_{e,SR}$ and $p_{e,RD}$ as the instantaneous SER of the first and second hops, respectively.

$$\begin{aligned}
p_e &= 1 - \left(1 - Q\left(\sqrt{2r_{SR}}\right)\right)\left(1 - Q\left(\sqrt{2r_{RD}}\right)\right) \\
&\leq Q\left(\sqrt{2r_{SR}}\right) + Q\left(\sqrt{2r_{RD}}\right)
\end{aligned} \quad (6)$$

Therefore, we can get the average SER of the first hop according to [5], [6] as

$$\bar{P}_{e,SR} = \frac{1}{2} - \sqrt{\frac{r_s\sigma_{SR}^2}{r\delta_{CR}^2}}exp\left(\frac{r_s\sigma_{SR}^2}{r\delta_{CR}^2}\right)\Gamma\left(-\frac{1}{2}, \frac{r_s\sigma_{SR}^2}{r\delta_{CR}^2}\right) \quad (7)$$

We get on average the conditional SER of the second hop according to [5], [6].

$$\begin{aligned}
\bar{P}_{e,RD} &= \frac{1}{2} - \sqrt{2}\sqrt{\frac{r_R\delta_{RD}^2}{r\delta_{CR}^2}}exp\left(\frac{r_R\delta_{RD}^2}{2r\delta_{CR}^2}\right)D_{-3}\left(\frac{2r_R\delta_{RD}^2}{r\delta_{CR}^2}\right) \\
&- 4r^2\left(\delta_{CR}^2\right)^2\sqrt{r\delta_{CR}^2 r_R\delta_{RD}^2}exp\left(\frac{r_R\delta_{RD}^2}{2r\delta_{CR}^2}\right)D_3\left(\frac{2r_R\delta_{RD}^2}{r\delta_{CR}^2}\right)
\end{aligned} \quad (8)$$

Proof: See Appendix B.

Then the average SER upper bound of the whole system can be expressed as

$$\begin{aligned}
\bar{P}_e &\leq \bar{P}_{e,SR} + \bar{P}_{e,RD} \\
&\leq 1 - \sqrt{\frac{r_s\sigma_{SR}^2}{r\delta_{CR}^2}}exp\left(\frac{r_s\sigma_{SR}^2}{r\delta_{CR}^2}\right)\Gamma\left(-\frac{1}{2}, \frac{r_s\sigma_{SR}^2}{r\delta_{CR}^2}\right) \\
&- \sqrt{2}\sqrt{\frac{r_R\delta_{RD}^2}{r\delta_{CR}^2}}exp\left(\frac{r_R\delta_{RD}^2}{2r\delta_{CR}^2}\right)D_{-3}\left(\frac{2r_R\delta_{RD}^2}{r\delta_{CR}^2}\right) \\
&- 4r^2\left(\delta_{CR}^2\right)^2\sqrt{r\delta_{CR}^2 r_R\delta_{RD}^2}exp\left(\frac{r_R\delta_{RD}^2}{2r\delta_{CR}^2}\right)D_3\left(\frac{2r_R\delta_{RD}^2}{r\delta_{CR}^2}\right)
\end{aligned} \quad (9)$$

4 SIMULATIONS

In this section, we give some numerical examples of the outage probability and the SER as a function of SNR. We assume one-antenna source node, one-antenna relay node and multi-antenna destination, which are located at (0, 0), (5, 5), and (10, 0), respectively. The distance is denoted by d. All the channels are subject to Rayleigh fading and uniform path-loss, with a loss exponent $\alpha = 4$.

Figure 1 shows the system outage probability when the number of antennas is changed. We compare the performance of the proposed multiple antennas with interference and single antennas with interference. We set $rth = 10$ dB. From Figure 1, we can see that the more antennas are deployed at the destination node, the probability of outage will occur less. The performance gets better as the ratio becomes larger.

Figure 2 depicts the SER performance. The SER of the whole system due to eqn. (9) is too complex, so the paper just analyses the average SER upper bound of the whole system. In this figure, we show the SER performance of the multiple antennas with interference, multiple antennas without interference

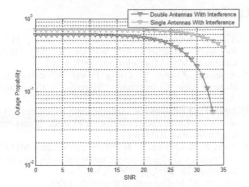

Figure 1. Comparison of outage probability.

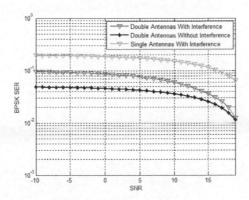

Figure 2. BPSK SER versus SNR.

and single antenna with interference. Comparing the performance of single and multiple antennas with interference, we can see that the SER performance of multiple antennas is obviously better than single antenna. The higher the SNR, the more obvious the performance. But it is also seen that the SER performance is smaller when comparing the SER performance of the multiple antennas with interference and multiple antennas without interference. The difference between the two is so obvious that we had better not ignore the interference.

5 SUMMARY

In this paper, we propose a regenerative relaying with multiple antennas at the destination node. We take the interference into consideration as the D2D pair and CU share the same UP resources. Numerical results show that our system outperforms than the system with single antenna from the perspective of outage probability and SER. The higher the SNR, the more obvious the performance. As the difference is obvious, we shall consider the interference between the D2D pair and CU.

ACKNOWLEDGEMENT

This work is supported by the Natural Science Foundation of China 61171093, the Important National Science and Technology Specific Project 2012ZX03003011-005, and the Scientific Research Fund of Nanjing University of Posts and Telecommunications NY211033.

APPENDIX

A. Evaluating $P(r_{SR} \geq rth)$ and $P(r_{RD} \geq rth)$

$$P(r_{SR} \geq rth) \approx P\left(\frac{r_s |h_{SR}|^2}{r|h_{CR}|^2} \geq rth\right)$$

$$= \int_{rth}^{\infty}\int_0^{\infty} \frac{y}{r_s\sigma_{SR}^2} exp\left(-\frac{yz}{r_s\sigma_{SR}^2}\right)$$

$$\times \frac{1}{r\sigma_{CR}^2} exp\left(-\frac{y}{r\sigma_{CR}^2}\right)dydz \text{ with}$$

the help of [6, eq. (2.322.2)]
we can get

$$= \frac{r_s\delta_{SR}^2}{rth^2 r\delta_{CR}^2 + r_s\delta_{SR}^2}$$

$$P(r_{RD} \geq rth) \approx P\left(\frac{r_R|h_{RD1}|^2 + r_R|h_{RD2}|^2}{r|h_{CD}|^2} \geq rth\right)$$

$$= \int_{rth}^{\infty}\int_0^{\infty}\left(\frac{1}{r_R\delta_{RD}^2}\right)^2$$

$$exp\left(-\frac{yz}{r_R\delta_{RD}^2} - \frac{y}{r\delta_{CR}^2}\right)\frac{y^2 z}{r\delta_{CR}^2}dydz$$

$$= \frac{r_R\delta_{RD}^2\left(2r\delta_{CR}^2 rth + r_R\delta_{RD}^2\right)}{\left(r\delta_{CR}^2 rth + r_R\delta_{RD}^2\right)^2}$$

B. Evaluating $p_{e,SR}$ and $p_{e,RD}$
According to A, we can evaluate $f(r_1)$ and $f(r_2)$

$$f(r_1) = \frac{1/krr_s\sigma_{SR}^2\sigma_{CR}^2}{\left(\dfrac{x}{r_s\sigma_{SR}^2} + \dfrac{1}{kr\sigma_{CR}^2}\right)^2} = \frac{A}{(Bx+C)^2}$$

$$f(r_2) = \frac{2xr_R\delta_{RD}^2 r^2\left(\delta_{CR}^2\right)^2}{\left(r\delta_{CR}^2 x + r_R\delta_{RD}^2\right)^3} = \frac{2\alpha x}{(\beta x + \varsigma)^3}$$

(7) can be evaluated with the help of [5, eq. (3.383.10)], [6, eq. (3.3213)].

$$\bar{p}_{e,SR} = \int_0^\infty \int_{\sqrt{2x}}^\infty \frac{1}{\sqrt{2\pi}} exp\left(-\frac{y^2}{2}\right) \frac{A}{(Bx+C)^2} dxdy$$

$$= \int_0^\infty \int_0^{\frac{y^2}{2}} \frac{1}{\sqrt{2\pi}} exp\left(-\frac{y^2}{2}\right) \frac{A}{(Bx+C)^2} dxdy$$

$$= \frac{1}{2} - \sqrt{\frac{r_s\sigma_{SR}^2}{r\delta_{CR}^2}} exp\left(\frac{r_s\sigma_{SR}^2}{r\delta_{CR}^2}\right) \Gamma\left(-\frac{1}{2}, \frac{r_s\sigma_{SR}^2}{r\delta_{CR}^2}\right)$$

According to [5, eq. (3.383.7)], [6, eq. (2.114)],

$$\bar{p}_{e,RD} = \int_0^\infty \int_{\sqrt{2x}}^\infty \frac{1}{\sqrt{2\pi}} exp\left(-\frac{y^2}{2}\right) \frac{2\alpha x}{(\beta x + \xi)^3} dxdy$$

$$= \int_0^\infty \int_0^{\frac{y^2}{2}} \frac{1}{\sqrt{2\pi}} exp\left(-\frac{y^2}{2}\right) \frac{2\alpha x}{(\beta x + \xi)^3} dxdy$$

$$= \frac{1}{2} - \sqrt{2}\sqrt{\frac{r_R\delta_{RD}^2}{r\delta_{CR}^2}} exp\left(\frac{r_R\delta_{RD}^2}{2r\delta_{CR}^2}\right) D_{-3}\left(\frac{2r_R\delta_{RD}^2}{r\delta_{CR}^2}\right)$$

$$- 4r^2\left(\delta_{CR}^2\right)^2 \sqrt{r\delta_{CR}^2 r_R\delta_{RD}^2} exp\left(\frac{r_R\delta_{RD}^2}{2r\delta_{CR}^2}\right)$$

$$\times D_3\left(\frac{2r_R\delta_{RD}^2}{r\delta_{CR}^2}\right)$$

REFERENCES

[1] Yingbin Liang, Veeravalli, V.V. and Poor, H.V., Resource Allocation for Wireless Fading Relay Channels: Max-Min Solution, 2007, pp.61.

[2] Xiaojuan Zhang, Yi Gong Power Allocation for Regenerative Cooperative Systems with Multi-Antenna Destination. 2008, pp.3755–3759.

[3] Janis, P., Koivunen, V., Interference-aware resource allocation for device-to-device radio underlaying cellular networks. 2009, pp.1–5.

[4] Janis, P., Koivunen, V., Interference-avoiding MIMO schemes for device-to-device radio underlaying cellular networks. 2009, pp.2385–2389.

[5] Knight. K, Mathematical Statistics. 1999.

[6] Gradshteyn I.S., Ryzhik I.M. Table of Integrals, Series, and Products. 2000.

Electrical, Control Engineering and Computer Science – Liu (Ed.)
© *2016 Taylor & Francis Group, London, ISBN 978-1-138-02937-8*

Channel estimation of IEEE 802.11ah based on traveling pilot and compressive sensing

Yue Wang, Yunzhou Li & Jungang Hu
Research Institute of Information Technology, Tsinghua University, Beijing, China

ABSTRACT: IEEE 802.11ah amendment proposes traveling pilot, which can be combined with Long Training Sequence (LTF) to refresh the initial channel estimation and to effectively cope with the Doppler shift. We analyzed the performance of traveling pilot via Compressive Sensing (CS) based on OMP algorithm, and compared it with LS estimate and MMSE estimate. The results show that, compared with other interpolation algorithms, OMP algorithm can fully utilize traveling pilot's scattered structure which gets the best overall performance.

Keywords: IEEE 802.11ah; compressive sensing; traveling pilot; channel estimation

1 INTRODUCTION

OFDM system has already been the major choice when we implement the physical layer of a wireless communication system. In order to ensure the stability and accuracy of communication, channel estimation is an essential part of an OFDM receiver. Typically, we improve channel estimation performance in two ways: pilot pattern design and channel estimation algorithm. In terms of pilot pattern design, Block-type pilot [1], Comb-type pilot [2], and Square-type pilot [3] are frequently used. We classify them to fixed pilot collectively. In terms of channel estimation algorithm, there are several common algorithms such as LS estimate [4], MMSE estimate [5], and its reduced form LMMSE estimate [6].

However, the problem of traditional 802.11 systems is that they only enable channel estimation at the beginning of the packets during the LTF while lack of the ability to continuous refresh the initial channel estimation throughout the packet. That is why we choose traveling pilots. On the other hand, although the complexity of LS estimate is low, it would be very sensitive to AWGN. MMSE estimate can guarantee accuracy, but its complexity is too high for implementation. Even though we can simplify the calculation process by SVD, its performance is still not satisfactorily enough. In addition, we need to know some channel properties such as the autocovariance matrix of channel transfer function h.

Therefore, we combine traveling pilot and OMP algorithm to analyze channel estimation for 802.11ah.

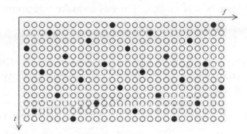

Figure 1. Pilot pattern when 32FFT, $N_{sts} = 1$.

2 STRUCTURE OF TRAVELING PILOT

According to [7], traveling pilot's function is when both transmitter and receiver are in outdoor environment and one of the paths is the reflection of a moving car, the transmission performance by using traveling pilot is superior to just using traditional fixed pilot. The reason is that the location of the pilot continually changes so that all tones are covered, thus enabling a continuous refresh of the initial channel estimation throughout the packet.

In 32FFT case, traveling pilot insertion position is shown in Figure 1, in which N_{sts} means the number of space–time streams. Other structures such as 64FFT, 128FFT, and 256FFT can be found in [8].

3 CHANNEL ESTIMATION BASED ON COMPRESSIVE SENSING

This chapter mainly focuses on three cores of compressive sensing: signal sparsity, measurement matrix design, and signal reconstruction algorithm [9].

Sparsity aims to find an orthogonal basis matrix Ψ, which can make the signal representation in the orthogonal basis sparse or approximately sparse.

Measurement matrix design aims to find a measurement matrix, which is steady and uncorrelated with the orthogonal basis so that the high dimensional signal can be reduced to low dimensional to get fewer samples.

Signal reconstruction algorithm focuses on how to design a fast and effective way to reconstruct original signal from the reduced samples.

3.1 Signal sparsity

Consider a discrete signal $x(n)$, $n \in [1, 2, ..., N]$

$$x = \Psi \alpha \tag{1}$$

According to Eq. 1, it can be represented by a linear combination of basis $\Psi^T = [\Psi_1, \Psi_2, ..., \Psi_N]$, where x and α are vectors with N dimensions, and Ψ is a N × N matrix. Ψ needs to make x K-sparse.

3.2 Measurement matrix design

The aim of measurement matrix design is to find a matrix Φ, which is M × N and uncorrelated with the basis Ψ.

$$y = \Phi x = \Phi \Psi \alpha = \Theta \alpha \tag{2}$$

where y is called observation matrix which is actually a vector with M dimensions.

The process of recovering original signal from y is called signal reconstruction based on compressive sensing. Candes claims that if Θ satisfies the Restricted Isometric Property (RIP) [10], shown in Eq. 3, signals still can be recovered by OMP algorithm.

$$(1-\delta_K)\|x\|_2^2 \leq \|\Theta x\|_2^2 \leq (1+\delta_K)\|x\|_2^2 \tag{3}$$

3.3 Signal reconstruction algorithm

The task of reconstruction algorithm is to recover signal x from observation matrix y.

First, we need to get the channel impulse response.

When the channel coherence time is much longer than the OFDM symbol period, it can be considered that channel impulse response is stable in one symbol period. So channel impulse response can be expressed as

$$h(t) = \sum_{l=0}^{L-1} a_l \delta(t - \tau_l T_s) \tag{4}$$

where T_s is the sample interval, L is τ_{max}/T_s rounding up which equals the number of taps in discrete-time channel model, and τ_{max} is the largest delay of paths.

Define DFT matrix as

$$F = \begin{pmatrix} W_N^{00} & \cdots & W_N^{0(N-1)} \\ \vdots & \ddots & \vdots \\ W_N^{(N-1)0} & \cdots & W_N^{(N-1)(N-1)} \end{pmatrix}_{N \times N}$$

where $W_N = \left(1/\sqrt{N}\right)e^{-j2\pi/n}$.

Meanwhile, define $Fh = DFT(h)$ as a channel frequency response, N_T as time domain noise, and N_F as frequency domain noise. Define $X = DFT(x)$, then we can derive the signal at receiver

$$Y = DFT(IDFT(X)*h) + N_T = XFh + N_F \tag{5}$$

Define $E = (e_{S_1}, e_{S_2}, ..., e_{S_n})$ as pilot selection matrix, which is used to extract pilots from signal Y at receiver. X is a N × N matrix. S_n means the pilot position in DFT matrix, e_{S_n} is a N × 1 vector, so S_n is N × N. And Fh is N × 1. After selection, response at pilot points is

$$Y_p = XEFh + N_{pF} \tag{6}$$

As the receiver knows original pilots sent by transmitter, the impulse response at the pilot points can be described as

$$\hat{H}_P = Y_p/X_p = EFh + N'_{pF} = \Phi h + N'_{pF} \tag{7}$$

where Φ is measurement matrix. As E and F have been known by receiver, Φ can also be calculated. Then, we derive channel response h.

The key procedures of OMP algorithm are as follows:

Aim: Calculate the position where the transvection of measurement matrix and residual is largest through K times' iteration.

Input: observation y, residual $r = y$, sparsity K and measurement matrix Φ.

Output: position index set Λ and corresponding value S.

Procedure:

1. Define i as current iteration number, set initial value: residual $r_0 = y$, index set $\Lambda = \varnothing$.
2. find the index λ_i which satisfies $\lambda_i = \text{argmax}_k |<r_{i-1}, \varphi_k>|$, where i is the current iteration number, and k is the number of column of Φ.
3. update $\Lambda_i = \Lambda_{i-1} \cup \lambda_i$, and add one column to Φ_{sub} from Φ according to Λ_i which can be described as $\Phi_{sub} = \Phi_{\Lambda_i}$.

4. solve $z = \arg\min \| Y - \Phi_{sub}S \| = \Phi_{sub}^{\dagger} Y$, by the least square method, † indicates pseudo-inverse
5. calculate residual $r_i = Y - \Phi_{sub}z$; if $i < K$, return to step (2); if $i = K$, go to next step
6. obtain an estimate for the original signal.

4 SIMULATION AND RESULTS

We performed computer simulations to evaluate the model. Our system uses 1 MHz bandwidth with 32 sub-carriers. Channel model is SISO multiple-path model: Pedestrian A which has Doppler shift on its fourth path [11]. The parameters of the model is as follows:

PDP		Pedestrian A	
# of paths		4	
Relative path power (dB)	Delay (ns)	0	0
		−9.7	110
		−19.2	190
		−22.8	410

We combined LTF (Long Training Field) and traveling pilot for channel estimation.

Figure 2 depicts the normalized MSE which is defined as

$$MSE = \frac{E \left\| \sum_i H(i) - \hat{H}(i) \right\|^2 \|}{E \left\| \sum_i H(i) \right\|^2 \|} \tag{8}$$

We simulated the performance of OMP, LS, and MMSE estimate under SNR ranging from 0 to 30 dB.

First, MSE of three algorithms are all reduced during the increase in the training sequence length. Meanwhile, the mean square error of OMP algorithm is about 6 dB less than that of LS estimate. Although the MSE of MMSE estimate is lower than OMP algorithm, the complexity is sacrificed. Setting the number of sub-carriers to N, the number of pilot to P, the signal length to L, and the sparsity to K, the total computation complexity of MMSE is $5N^4 + N^3 + 6N^2 - N$ [12]. The complexity of OMP algorithm is $O(LK^2)$. The number of iteration D in the OMP is approximately equal to the sparsity K. Even though we have to make a pseudo-inverse operation during each iteration, the number of pseudo-inverse operation just increases from 1 to D which will not be too large. If some fast pseudo-inverse computing methods can be used, the complexity will be lower. The LS estimate requires two times of inverse operation of P-order matrix, and its complexity is $O(P^3)$. Because the number of pilots is much larger than the sparsity, when we use few pilots, the complexity of LS and OMP are approximately the same; when we use massive pilots, the complexity of LS estimate is much higher than OMP algorithm. In conclusion, the performance of OMP algorithm is the best.

Figure 3 shows the BER performance of OMP algorithm and LS estimate. It is seen that the BER performance of OMP is much better than LS estimate. We know that the ratio of traveling pilot in OFDM symbols is pretty low. It can be only 1/13, for example, in 32FFT. Therefore, compared with OMP algorithm which uses sparse matrix to accurately recover part of impulse response in low SNR, LS estimate has higher BER due to deficiency of noise reduction and interpolation. Meanwhile, the

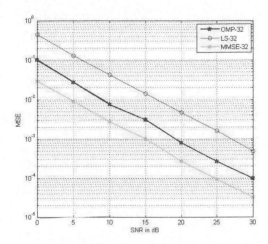

Figure 2. MSE performance comparison between OMP, LS, and MMSE.

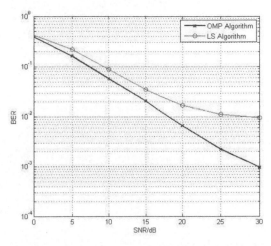

Figure 3. BER performance comparison between OMP and LS.

BER performance of LS estimate has had the trend to be flat when SNR is 30 dB.

5 SUMMARY

In this paper, we mainly discuss channel estimation methods for 802.11ah protocol. We present a channel estimation method which combines compressive sensing and traveling pilot so as to utilize their own advantages. The simulation result shows that, compared with other interpolation algorithms, OMP algorithm can utilize the advantage of traveling pilot pattern effectively which gets the best performance.

REFERENCES

[1] Y. Zhao and A. Huang. "A Novel Channel Estimation Method for OFDM Mobile Communication Systems Based on Pilot Signals and Transform-Domain Processing." Proceeding of IEEE VTC 47th Vehicular Technology Conference, Vol. 3, pp. 2089–2093, 5–7, May, 1997.

[2] M.H. Hsieh, C.H. Wei. "Channel Estimation for OFDM System Based on Comb-type Pilot Arrangement in Frequency Selective Fading Channels." IEEE Transactions on Consumer Electronics, Vol. 44, No. 1, Feb. 1998.

[3] P. Hoeher, S. Kaiser and P. Robertson. "Two-dimensional pilot-symbol-aided channel estimation by Wiener filtering." In Proc. 1997 IEEE Int. Conf. Acoustics, Speech and Signal Processing, Munich, Germany, Apr. 1997, pp. 1845–1848.

[4] J.J. van de Beek, O. Edfors, M. Sandell, S.K. Wilson, and P.O. Borjesson. "On channel estimation in OFDM systems." In Proc. 45th IEEE Vehicular Technology Conf., Chicago, IL, July 1995, pp. 815–819.

[5] Senol, H., Cirpan, H.A., Panayirci E., "Pilot-aided Bayesian MMSE channel estimation for OFDM systems: algorithm and performance analysis." IEEE Global Telecommunications Conference, Vol.4 Dec. 2004, pp. 2361–2365.

[6] O. Edfors, M. Sandell, J.J. van de Beek, S.K. Wilson, and P. O. Borjesson. "OFDM channel estimation by singular value decomposition." IEEE Trans. Commun., vol. 46, pp. 931–939, Jul. 1998.

[7] IEEE 802.11ah. 11–12/1322r0 11ah traveling-pilots [C]. USA: 2012.11.

[8] IEEE 802.11ah. 11/1137 11ah Specification Framework for TGah [C]. USA: 2013.5.

[9] Berger, C.R.; Carnegie Mellon Univ., Pittsburgh, PA, USA; Zhaohui Wang; Jianzhong Huang; Shengli Zhou. Application of Compressive Sensing to Sparse Channel Estimation [J]. Communications Magazine, IEEE, 2010, (48): 164–174.

[10] E.J. Candes, J. Romberg and T. Tao, "Robust Uncertainty Principles: Exact Signal Reconstruction from Highly Incomplete Frequency Information," Information Theory, IEEE Transactions on, Vol. 52, No. 2, pp. 489–509.

[11] IEEE 802.11ah 11/0760 TGah Outdoor Channel Models—Revised Text [C] UCA: 2011.5.

[12] Wu Jun-qin, Zhao Xue. LTE Downlink Performance Analysis of Two-Dimensional Channel Estimation Algorithm Based on Time-Frequency [J]. Computer Engineering and Design. 2013, (34): 1162–1166.

Electrical, Control Engineering and Computer Science – Liu (Ed.)
© 2016 Taylor & Francis Group, London, ISBN 978-1-138-02937-8

Design and implementation of embedded network communication system based on RT-thread and LwIP

L.F. Huang, N.G. Chen, L.Y. Huang & H.Z. Lin
College of Information Science and Technology, Xiamen University, Xiamen, China

ABSTRACT: This paper details the method of connecting the embedded development board Nuvoton N32926 to the Ethernet using the LwIP protocol stack based on the RT-Thread system. It focuses on analysis and implementation of the N32926's network control and the design and implementation of the network card RTL8201 driver, and then it tests the Network Communication function of the system. The test results show that the embedded devices have the ability of network communication.

Keywords: Nuvoton N32926; RT-thread; LwIP; RTL8201

1 INTRODUCTION

With the development of network and technology based on embedded system, an increasing number of embedded devices are required to achieve the function to communicate with Internet. The problem lies in the realization of TCP/IP stack of embedded devices. Although the complete TCP/IP is powerful, it is too huge to be transplanted to the embedded devices. The TCP/IP protocol of embedded devices must be concise, transplantable, and reducible.

LwIP (A Lightweight TCP/IP Protocol) is an embedded TCP/IP stack with advantages of full functionality, free code, and can be transplanted easily. It occupies only around tens of kb of RAM and 40 kb of ROM, which makes it a perfect choice for embedded devices.

RT-Thread is a widely used embedded operating system with LwIP as the TCP/IP protocol stack. Based on the RT-Thread transplanted in Nuvoton N32926 development board, this paper describes the method of designing the network card driver and providing the ability of network communication for system with the LwIP protocol stack. And finally the system is proved capable to communicate with Internet.

2 SYSTEM PLATFORM

The hardware platform of the system adopts Nuvoton N32926. The N32926 development board is built on the ARM926EJ-S CPU core and integrated with network card RTL8201F, which meets the physical layer's need while sending and receiving network data. It also provides an Ethernet MAC Controller (EMC) for WAN/LAN application. This EMC has its DMA controller, transmission FIFO, and reception FIFO.

The software environment is the RT-Thread operating system. RT-Thread is a real-time operating system with open source code. It is designed at object-oriented style and implemented with C language. Its outstanding features include small volume, efficient, and can be cut easily. The core is so small that it occupies only 2.5 kb of ROM and 1 kb RAM space. Overall system block diagram is shown in Figure 1.

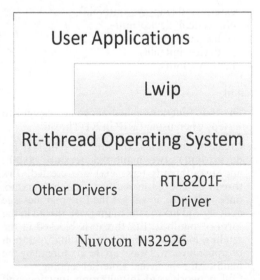

Figure 1. System block diagram.

3 LWIP IN RT-THREAD

RT-Thread uses an independent double-thread structure for the Ethernet data transceiving. LwIP protocol stack in the system runs as follows.

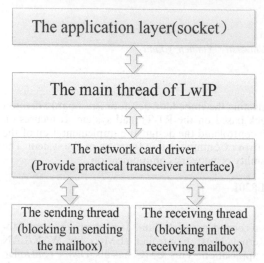

Figure 2. The overall framework of LwIP operation.

Function rt_application_init () is the entrance of user program in RT-Thread system, it initializes the user's application threads, and once the scheduler is opened, user thread will be implemented. It also starts the init_thread thread where LwIP protocol stack is called as follows:

```
#ifdef RT_USING_LWIP
    {
        extern void lwip_sys_init(void);
        eth_system_device_init();
        rt_hw_emac_init();
        rt_device_init_all();
        lwip_sys_init();
    }
#endif
```

The four functions called above correspond to core steps when using LwIP in RT-Thread:

1. Create transceiving threads. In eth_system_device_init(), two mailboxes (erxmb, etxmb) and two threads (erx, etx) are created. Erx thread is blocked in receiving mailbox erxmb, once waked up, it will get the Ethernet message actively and inform the main LwIP thread in the form of mailbox. Etx thread is blocked in the sending mailbox etxmb, the sending operation of upper layer will wake up the etx thread by the mailbox for the actual Ethernet sending.
2. Call network card initialization function and register network card device. The interface

rt_hw_emac_init() initializes the network device, and function of rt_device_init_all () is to initialize all devices that are registered in object manager, but have not been initialized yet.

3. Initialize LwIP and create LwIP main thread. This part is accomplished in lwip_sys_init () interface.

4 DESIGN AND IMPLEMENTATION OF NETWORK CARD DRIVER

In RT-Thread, the device is considered to be a kind of object. Each device object derives from the base object. That means device can inherit the properties of his father class objects and derives its private properties. The RT-Thread system provides a simple I/O device management framework, and upper layer applications get the right device driver through the device operation interface and finish the data interaction with the underlying hardware with the device driver.

The RTL8201F network card driver is responsible for device initialization and data transmission and reception. The following table is the main interface it provides to the upper layer.

First of all, a new device structure inherited from struct eth_device needs to be defined.

Figure 3. Equipment operation interface and its driver.

Table 1. Interfaces RTL8201F driver provides.

Function	Interfaces driver provides
Device initialization	static rt_err_t rt_emac_init(rt_device_t dev)
Device close	static rt_err_t rt_emac_close(rt_device_t dev)
Receiving data	struct pbuf* rt_emac_rx(rt_device_t dev)
Sending data	rt_err_t rt_emac_tx(rt_device_t dev, struct pbuf* p)
Network driver interface	void rt_hw_emac_init()
Card receiving interrupt processing	void MAC0_Rx_isr(int vector, void *param)

The structure emac_device in the RTL8201F driver is defined as follows:

struct rt_macb_eth
{
    struct eth_device parent;
    rt_uint8_t dev_addr[MAX_ADDR_LEN];
};

rt_emac_init (rt_device_t dev) is a device initialization function. RTL8201F is initialized after the CPU configuration of internal Network Control Register (NCR) and interrupt registers. The initialization process includes software reset, setting network working mode, setting the PHY, selecting mode, enabling the RX/TX interrupt, enabling data receiving function and so on. And in rt_emac_close (rt_device_t dev), function EMAC_Exit () is called to close the device.

LwIP supports many network interfaces. Each network interface corresponds to a struct netif, which contains transceiver functions corresponding to the network interface. The LwIP receives and sends Ethernet packets by calling netif->input () and netif->output (), while rt_emac_rx () and rt_emac_tx () are the actual interfaces it finally calls. Rt_emac_rx () is responsible for receiving Ethernet data and packing it into structure that LwIP can recognize. The main purpose of rt_emac_tx() is to pack the data to be sent into the structure that network card can recognize and then send the data, this function realizes the real process of packet transmission.

RT-Thread maintains a linked list for all devices. A device has to register to the list before it can be visited by the upper application. Therefore, eth_device_init() should be called in the network driver interface rt_hw_emac_init () to initialize the network card device, add the name, IP, subnet mask, gateway, transceiver functions for the network card, and finally register the network card to the system.

The receiving thread erx blocks in mailbox erxmb, when the Ethernet hardware device receives network packets and generates interrupt, interrupt service MAC0_Rx_isr () should call eth_device_ready () to send mail to erxmb mailbox to wake up erx thread, then erx thread will get the Ethernet packets actively.

5 TEST

The development board IP address is set to 192.168.1.170 and connected to the same network segment with PC whose IP address is 192.168.1.108. First of all, ping the IP address 192.168.1.170 in the PC machine, it turns out that the packets PC sends are all received, as shown in Figure 4. The average of return time is 0 ms and the packet loss rate is 0,

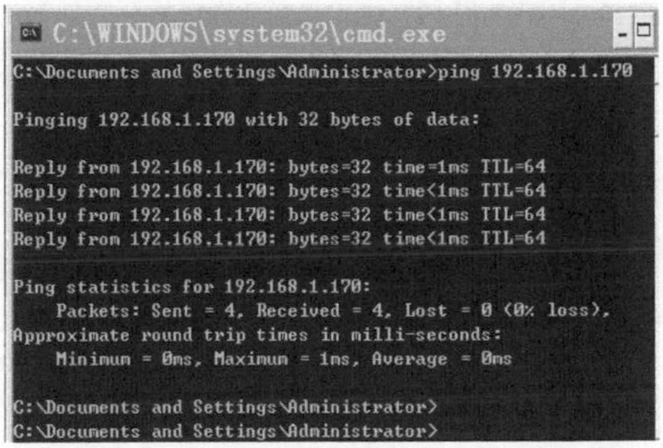

Figure 4. Result of ping test.

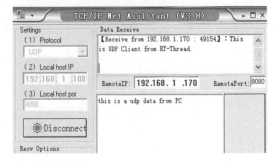

Figure 5. UDP Server and client test.

![Figure 6 screenshot of TCP/IP Net Assistant]

Figure 6. Result of test in net assistant.

which means the LwIP protocol stack has been run successfully. Second, start UDPClient thread in RT-Thread and send data "this is UDP Client from RT-Thread" to socket 192.168.1.108:8080. The data are received correctly by PC, which proves that the embedded development platform has the ability of sending data to Internet. Lastly, start UDP Server thread in RT-Thtread and waiting for client on port 8080. Using software Net Assistant, PC sends data "this is an udp data from

PC" to socket 192.168.1.170:8080. The data are also received correctly; results of tests are shown in Figures 5 and 6. So far, send and receive abilities of development have all been proved.

6 CONCLUSION

This paper discusses how to realize the function of network communication in ARM9 embedded system using RT-Thread. It analyzes the mechanism of LwIP module in the system and puts emphasis on realization of RTL8201F network card driver. Finally, it proves that the embedded equipment accesses Ethernet successfully by some tests.

REFERENCES

J. Zhang, and M. Fu, "The porting and application of the internet protocol lwip on the embedded Linux system", Microcomputer Information, 2011, (04):94–96.

X.C. Chen, and J. Wang, "Design and Implementation of Embedded Network Communication System Based on RT—Thread", Audio Engineering, 2012, (12):49–56.

Z.Y. Wang, S.G. Yang, J.W. Wang and Z.H. Deng, "Transplant and Test of Embedded LwIP Stack", Computer & Digital Engineering, 2014, (02):272–275.

C.H. Fu, and S. Yang, "Study on the Porting and Application of LwIP Based on uC/OS-II and ARM", Journal of Southwest University of Science and Technology, 2009, (03):71–74.

D. Kong, and J.H. Zheng, "Transplant and Application of LWIP in ARM Platform", Communications Technology, 2008, (06):38–40.

Electrical, Control Engineering and Computer Science – Liu (Ed.)
© 2016 Taylor & Francis Group, London, ISBN 978-1-138-02937-8

The hardware decoding implementation of video surveillance based on Android system

S.K. Liu, L. Cao, Z.Y. Shi & C. Feng
College of Information Science and Technology, Xiamen University, Xiamen, China

ABSTRACT: In order to achieve mobile video surveillance, this paper proposes a video surveillance system based on an Android phone. The paper takes Android system as the platform, presents a real-time communication solution of a high quality video. Android application calls the data transmission library and video codec library through the JNI interface and uses the hardware decoding technology to decode the video. The decoding efficiency has been greatly improved, and the video becomes much clear and smoother. This paper tested the video surveillance terminal in the WLAN (Wireless Local Area Network) environment to achieve the goal of using Android phone for mobile video surveillance.

Keywords: hardware decoding; video surveillance; Android system; WLAN

1 INTRODUCTION

Nowadays, the demand of family users for home surveillance and security is increasing day by day. As an important part of safety protection, video surveillance system plays an important role in the security assurance and crime prevention. Now, users put forward higher requirements on the quality of video. They tend to be more high definition video, and more high processing capacity for terminal equipment. Traditional video surveillance is realized by software decoding technology based on the CPU to decode the video, and it can only play a low resolution video. The demand of resolution and the decoding computation for the video is higher which greatly consume the CPU operation ability. This leads to that the video doesn't play smooth. It has failed to meet people's vision requirements. With the development of the hardware chip technology, a lot of processor chips have gained hardware decoding capability. Hardware decoding technology significantly reduces the CPU usage and improves the video decoding speed. Therefore, the CPU can be released from the heavy video decoding operation, and the equipment has the capability of smoothly playing HD video.

This paper introduces a kind of video surveillance system based on Android phone terminal, which combines video hardware decoding technology with mobile multimedia technology, truly realizing mobile video surveillance. Among them, H.264 standard with its outstanding coding efficiency, network adaptability and error recovery ability has become a mainstream video coding format. In addition, the popularity of Android phone together with high-speed mobile broadband provides a platform. On this platform we can develop a lot of applications, which makes the use of Android phone to watch surveillance video possible.

2 THE INTRODUCTION OF ANDROID DEVELOPMENT

2.1 Android system architecture

Android system uses integrated strategy ideas, its operating system architecture is divided into four layers. From top to bottom consequence, they are the application, the application framework, core class libraries, and the Linux kernel, which also includes the Android operating environment in the third layer.

2.1.1 Application layer
Application is a program written in the Java language running in a virtual machine. Android itself contains some of the core application, and developers can also develop more distinctive Android applications on this basis.

2.1.2 Application framework layer
Android application development is based on the framework and components. In the layer, developers have all the permissions to access the API framework, thus simplifying the application development architecture design and improving the efficiency of application development greatly.

2.1.3 Core class libraries layer

Core class libraries include the standard C library, media function library, 2D and 3D graphics library, the browser engine and the SQLite engine. The developers who use these functions find it convenient to develop the application program interface and shorten the development process greatly.

2.1.4 Linux kernel layer

Android platform is based on the optimized Linux version 2.6.23 kernel to develop. The kernel layer is located between the hardware and the software layer as a virtual intermediate layer used to provide system services.

2.2 Android NDK and JNI technology

Android NDK is a set of integrated tool components based on C/C++ bottom development of API, designed to help developers to quickly develop C/C++ dynamic library. Developers simply need to modify the Makefile to create the dynamic library, which greatly reduces the packing work. In addition, the native C/C++ file compiled by the Android NDK cannot be used directly in Java; it is required by the Java code embedded in the native C/C++ — using the native code through the JNI (Java Native Interface).

2.3 H.264 standard

In the network video surveillance system, video compression is very necessary and important work. The network transmission system is unable to bear the massive data without compression. H.264 is currently the most advanced video compression algorithm, which consists of video coding layer VCL and the network abstraction layer NAL. The H.264 standard has improved coding efficiency and image quality and adapted to various network environments. It is very suitable for the requirement of mobile video surveillance of high compression rate and complex network environment.

3 THE FRAMEWORK OF VIDEO SURVEILLANCE SYSTEM

The system is mainly to achieve video surveillance of the Android mobile phone through the WLAN. Video surveillance system is composed of collection terminal, router and Android client. All of them are connected to the WLAN via WIFI, as is shown in Figure 1. The collection terminal is responsible for collecting video data, H.264 image compression and RTP (Real Time Protocol) slices packaged; through the router sends H.264 video

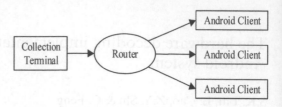

Figure 1. System framework.

data packets to the Android client; the Android client needs to provide the intuitive video operation interface for users, mainly to complete the H.264 video data decoding and display.

4 VIDEO HARDWARE DECODING IMPLEMENTATION

4.1 Multithreaded design

Receiving and decoding video data are a complex and ongoing process. The system is affected when one process is blocked. Therefore the client needs to use multiple threads to achieve the parallel processing of the data receiving and video decoding. In the whole process of running program, the main thread is responsible for screen refresh work in response to user operation, and creates two sub threads: data receiving and video decoding. The process is shown in Figure 2.

4.2 Hardware decoder

At present, Android system does not provide any corresponding API interfaces to achieve hardware decoding of the H.264 stream. While the software decoding method will lower the efficiency of image processing. This paper achieves the Android H.264 hardware decoder by using the libstagefrigt dynamic library which is in the Android libraries layer. The H.264 hardware decoder is programmed in C language which need to be compiled to dynamic library by the Android NDK. Then the JAVA code of the Android upper layer calls the dynamic library by using the JNI technology.

4.3 Decoding process

4.3.1 Initialization

The first step initializes the decoder and sets the relevant information of components by the decode_init function before decoding, such as the decoder type, the PPS and SPS parameters of the bit stream, the resolution of the image which is to be decoded and so on. The second step creates a hardware decoder by the Create function and finds

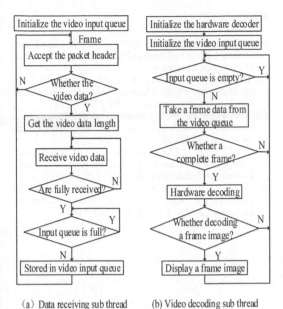

(a) Data receiving sub thread   (b) Video decoding sub thread

Figure 2. Sub-thread processing graph.

the related hardware decoder by the find CString function. OMXClient is a client that requests OMX IL for decoding work. If the decoder is created successfully, it will connect to the OMXClient. The third step creates an input queue. An input queue is opened by the in_queue function during initialization. Then the video stream to be decoded is sent into the input queue waiting for decoding.

4.3.2 Decoding

After the initialization of the hardware decoder, the decode_frame function is used to start the video decoding operation. Firstly, the video data that will be decoded in the Frame buffer is copied to the indummy_buf buffer through the memcpy function. After that, the video data is sent to the input queue to wait for decoding which will trigger a decoding thread. When the decoding thread is open, it will monitor whether the data is sent into the input queue. If there is data passed in, hardware decoder will be called for video decoding. After that, the decoded video data is sent into the buffer and ready to display.

4.3.3 Display

Decoded video data needs to be rendered before display. A rendering interface mVideoRenderer is created in the render_init function. Then decoded video data is sent into the mVideoBuffer to be rendered by the mVideoRenderer function. Finally, the mNativeWindow function is used to play the rendered video data. The rendered

video data will be played by the mNativeWindow function.

5 EXPERIMENTAL ANALYSES

Currently, the system has been tested in the laboratory. Android client terminal is installed on Android phone, through the WIFI access to the WLAN to establish a connection with the collection terminal. Android phone receives 30fps VGA format H.264 video data. The traditional software decoding of the video display interface is shown in Figure 3 and the hardware decoding of the video display interface is shown in Figure 4. From these two figures, we can know that the hardware

Figure 3. Display interface.

Figure 4. Display interface.

197

decoding technology improves the decoding efficiency and makes video images smoother, realizing the real-time video surveillance for Android phone.

6 CONCLUSIONS

According to video decoding performance requirements of the video surveillance system, this paper has implemented the video hardware decoding technology in Android system. Finally, the software was successfully running in the Android platform. It proved that this system can be used for video surveillance in the Android terminal, which has great practical value and application prospect.

REFERENCES

Han, C. & Liang, Q. 2010. The Principle and Main Points of Development of Android System. Publishing house of electronics industry.

Li, L.T. & Shi, Q.M. 2011. Research of Android Intelligent Mobile Phone Operating System. Science and technology information (25): 80–12.

Li, Y. & Feng, G. 2011. Development and Research on Multimedia Application Based on Android. Computer and modernization (4):149–152.

Zhang, G.Y. 2011. The Implementation of H.264 Video Compression Technology Based on Android System. Electronics technology 24(9):117–120.

Zhao, H.W. 2010. The Realization and Application of Development Environment of NDK. Computer knowledge and technology (35):119–120.

Author index

T - #0286 - 101024 - C0 - 246/174/10 [12] - CB - 9781138029378 - Gloss Lamination